Truly Nonlinear Oscillations

Harmonic Balance, Parameter Expansions, Iteration, and Averaging Methods

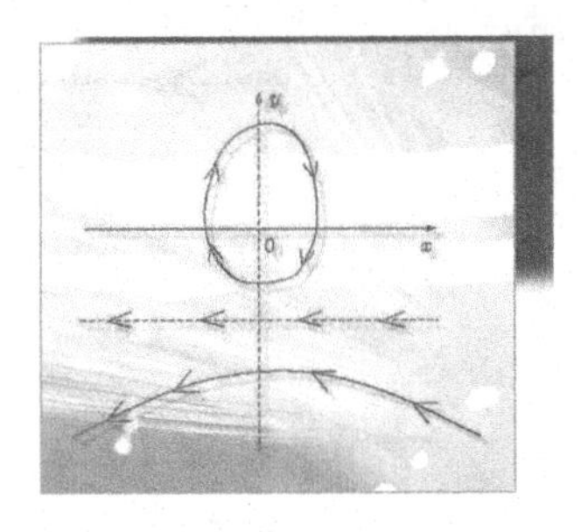

Truly Nonlinear Oscillations

Harmonic Balance, Parameter Expansions, Iteration, and Averaging Methods

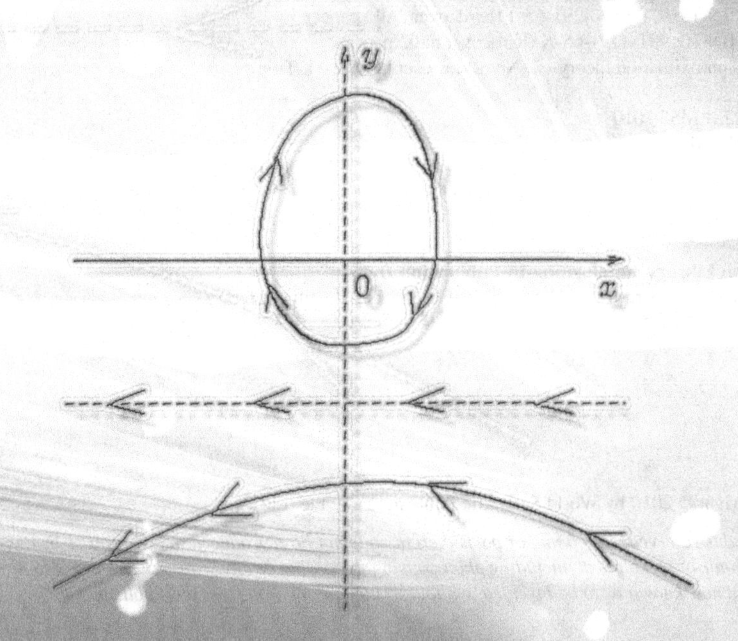

Ronald E Mickens

Clark Atlanta University, USA

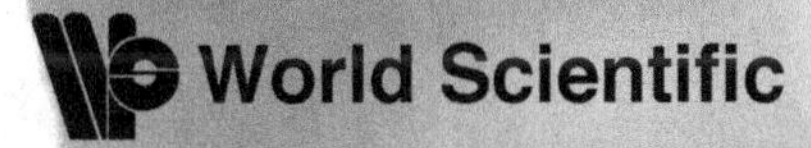

NEW JERSEY • LONDON • SINGAPORE • BEIJING • SHANGHAI • HONG KONG • TAIPEI • CHENNAI

Published by

World Scientific Publishing Co. Pte. Ltd.
5 Toh Tuck Link, Singapore 596224
USA office: 27 Warren Street, Suite 401-402, Hackensack, NJ 07601
UK office: 57 Shelton Street, Covent Garden, London WC2H 9HE

Library of Congress Cataloging-in-Publication Data
Mickens, Ronald E., 1943–
Truly nonlinear oscillations : harmonic balance, parameter expansions, iteration, and averaging methods / by Ronald E. Mickens.
p. cm.
Includes bibliographical references and index.
ISBN-13: 978-981-4291-65-1 (hardcover : alk. paper)
ISBN-10: 981-4291-65-X (hardcover : alk. paper)
1. Approximation theory. 2. Nonlinear oscillations. I. Title.

QA221.M53 2010
511'.4--dc22

2009038794

British Library Cataloguing-in-Publication Data
A catalogue record for this book is available from the British Library.

Printed in Singapore.

This book is dedicated to my family:
Daughter ... Leah,
Son ... James,
Wife ... Maria.

Preface

This small volume introduces several important methods for calculating approximations to the periodic solutions of "truly nonlinear" (TNL) oscillator differential equations. This class of equations take the form

$$\ddot{x} + g(x) = \epsilon F(x, \dot{x}),$$

where $g(x)$ has no linear approximation at $x = 0$. During the past several decades a broad range of calculational procedures for solving such differential equations have been created by an internationally based group of researchers. These techniques appear under headings such as

- averaging
- combined and linearization
- harmonic balance
- homotopy perturbation
- iteration
- parameter expansion
- variational iteration methods.

Further, these methodologies have not only been applied to TNL oscillators, but also to strongly nonlinear oscillations where a parameter may take on large values. Most of these techniques have undergone Darwinian type evolution and, as a consequence, a large number of papers are published each year on specializations of a particular method. While we have been thorough in our personal examination of the research literature, only those papers having an immediate connection to the topic under discussion are cited because of the magnitude of existing publications and because an interested user of this volume can easily locate the relevant materials from various websites.

We have written this book for the individual who wishes to learn, understand, and apply available techniques for analyzing and solving problems involving TNL oscillations. It is assumed that the reader of this volume has a background preparation that includes knowledge of perturbation methods for the standard oscillatory systems modeled by the equation

$$\ddot{x} + x = \epsilon F(x, \dot{x}).$$

In particular, this includes an understanding of concepts such as secular terms, limit-cycles, uniformly-valid approximations, and the elements of Fourier series.

The basic style and presentation of the material in this book is heuristic rather than rigorous. The references at the end of each chapter, along with an examination of relevant websites, will allow the reader to fully comprehend what is currently known about a particular technique. However, the reader should also realize that the creation and development of most of the methods discussed in this book do not derive from rigorous mathematical derivations. This task is a future project for those who have the interests and necessary background to carry out these procedures. However, these efforts are clearly not relevant for our present needs.

The book consists of seven chapters and several appendices. Chapter 1 offers an overview of the book. In particular, it presents a definition of TNL equations, introduces the concept of odd-parity systems, and calculates the exact solutions to four TNL oscillatory systems.

Chapter 2 provides a brief discussion of several procedures for a priori determining whether a given TNL differential equation has periodic and/or oscillatory solutions. The next four chapters present introductions to most of the significant procedures for calculating analytical approximations to the solutions of TNL differential equations. These chapters discuss, respectively, harmonic balance, parameter expansion, iteration, and averaging methods. Each chapter gives not only the basic methodology for each technique, but also provides a range of worked examples illustrating their use.

The last chapter considers six TNL oscillator equations and compares results obtained by all the methods that are applicable to each. It ends with general comments on TNL oscillators and provides a short listing of unresolved research problems.

We also include a number of appendices covering topics relevant to understanding the general issues covered in this book. The topics discussed range from certain mathematical relations to basic results on linear second-order differential equations having constant coefficients. Brief presentations are given on Fourier series, the Lindstedt-Poincaré perturbation method, and the standard first-order method of averaging. A final appendix, "Discrete Models of Two TNL Oscillators," illustrates the complexities that may arise when one attempts to construct discretizations to calculate numerical solutions.

I thank my many colleagues around the world for the interest in my work, their generalization of these results and their own original "creations" on the subject of TNL oscillations. As always, I am truly grateful to Ms. Annette Rohrs for her technical services in seeing that my handwritten pages were transformed into the present format. Both she and my wife, Maria Mickens, provided valuable editorial assistance and the needed encouragement to successfully complete this project. Finally, I wish to acknowledge Dr. Shirley Williams-Kirksey, Dean of the School of Arts and Sciences, for providing Professional Development Funds to assist in the completion of this project. Without this support the writing effort would not have been done on time.

Ronald E. Mickens
Atlanta, GA
August 2009

Contents

List of Figures

List of Tables

Chapter 1

Background and General Comments

This chapter introduces the basic, but fundamental concepts relating to the class of oscillators we call "truly nonlinear." The two phrases "truly nonlinear oscillators" and "truly nonlinear differential equations" are used interchangeable. In Sections 1.1 and 1.2, respectively, we define truly nonlinear (TNL) functions and TNL oscillators. Section 1.3 presents general comments regarding time reversal invariant systems and odd parity oscillators. Section 1.4 discusses the important topic of the elimination of dimensional quantities in the physical nonlinear differential equations through the use of scaling parameters. The existence of and exact solutions to four TNL oscillators are given in Section 1.5; this is followed by a brief overview of four methods that can be used to construct analytic approximations to the periodic solutions for TNL oscillator differential equations. We conclude the chapter with a set of possible criteria that may be used to judge the value of a calculational method for generating approximate solutions.

1.1 Truly Nonlinear Functions

A TNL function is defined with respect to its properties in a neighborhood at a given point. For our purposes, we select $x = 0$. Thus, for a function $f(x)$, we make the following definition:

Definition 1.1. $f(x)$ is a TNL function, at $x = 0$, if $f(x)$ has no linear approximation in any neighborhood of $x = 0$.

The following are several explicit examples of TNL functions

$$f_1(x) = x^3, \quad f_2(x) = x^{1/3}, \quad f_3(x) = x + x^{1/3}. \tag{1.1.1}$$

Note that each of these functions is defined for all real values of x, i.e., $-\infty < x < \infty$. Inspection of $f_1(x)$ and $f_2(x)$ clearly illustrate why they are TNL functions; however, the third function requires just a little more analysis to understand that it is a TNL function. We have

$$\frac{x}{x^{1/3}} = x^{2/3} \Rightarrow |x| < |x^{1/3}|, \quad 0 < |x| < 1, \tag{1.1.2}$$

and thus it follows that in a neighborhood of $x = 0$, the $x^{1/3}$ term dominates x.

Other examples of TNL functions include

$$\begin{cases} f_4(x) = \dfrac{1}{x}, \quad f_5(x) = \dfrac{1}{x^{3/5}}, \quad f_6(x) = \dfrac{x^3}{1+x^2}, \\ f_7(x) = |x|x, \quad f_8(x) + \dfrac{1}{x^{1/3}}. \end{cases} \tag{1.1.3}$$

In all of the above expressions, we have set possible constants, which could appear, equal to one. This does not change in any way the essential features of these functions.

1.2 Truly Nonlinear Oscillators

In this volume, we consider only one-degree-of-freedom systems that can be mathematically modeled by differential equations having (in the simplest case) the generic form

$$\ddot{x} + f(x) = 0, \tag{1.2.1}$$

where the "dots" denote time derivatives, i.e., $\dot{x} \equiv dx/dt$ and $\ddot{x} \equiv d^2x/dt^2$.

Definition 1.2. If $f(x)$ is a TNL function, then the second-order differential equation, as given in Eq. (1.2.1), is a TNL oscillator.

Using specific representations of $f(x)$, from Section 1.1, the following are particular examples of TNL oscillators:

$$\begin{cases} \ddot{x} + x^3 = 0, \\ \ddot{x} + x^{1/3} = 0, \\ \ddot{x} + x + x^{1/3} = 0, \\ \ddot{x} + \dfrac{x^3}{1+x^2} = 0, \\ \ddot{x} + |x|x = 0, \\ \ddot{x} + \dfrac{1}{x} = 0. \end{cases} \tag{1.2.2}$$

The TNL oscillator concept can also be extended to limit-cycle systems. Starting with the well-known van der Pol oscillator [1–4]

$$\ddot{x} + x = \epsilon(1 - x^2)\dot{x}, \tag{1.2.3}$$

where ϵ is a positive parameter, then the following TNL oscillator generalizations can be constructed [5, 6].

$$\begin{cases} \ddot{x} + x^3 = \epsilon(1 - x^2)\dot{x}, \\ \ddot{x} + x^{1/3} = \epsilon(1 - x^2)\dot{x}. \end{cases} \tag{1.2.4}$$

1.3 General Remarks

Let's consider in more detail the specifics of the structural properties of Eq. (1.2.1). In fact, we can also consider the more general form

$$\ddot{x} + F(x, \dot{x}) = 0, \tag{1.3.1}$$

where $F(x, \dot{x})$ depends on both x and its first derivative, $\dot{x}$.

When $F(x, \dot{x})$ depends only on x, then $F(x, \dot{x}) = f(x)$, and Eq. (1.3.1) becomes

$$\ddot{x} + f(x) = 0. \tag{1.3.2}$$

Defining $y \equiv \dot{x}$, a first-integral can be obtained [7] using

$$\frac{d^2x}{dt^2} = \frac{dy}{dt} = y\frac{dy}{dx}. \tag{1.3.3}$$

With this result, Eq. (1.3.2) can be integrated to give

$$\frac{y^2}{2} + V(x) = V(A), \tag{1.3.4}$$

where initial conditions, $x(0) = A$, $\dot{x}(0) = y(0) = 0$, were used, and $V(x)$ is the potential energy [8]. Within the domain of physics, this first integral is the total energy and the nonlinear oscillator is called a *conservative oscillator* [3, 8]. Note that this is a general result, not depending as to whether the oscillator is TNL or of the usual type. An important feature of the solutions for conservative oscillators is that they have periodic solutions that range over a continuous interval of initial values [7, 8].

We can also consider "generalized" conservative oscillators. These oscillators satisfy the condition $F(x, -\dot{x}) = F(x, \dot{x})$. A particular example is

$$\ddot{x} + (1 + \dot{x}^2)x^{1/3} = 0, \tag{1.3.5}$$

which is a TNL oscillator that can be rewritten to the form

$$y\frac{dy}{dx} + (1+y^2)x^{1/3} = 0. \tag{1.3.6}$$

Integrating gives

$$T(y) + V(x) = V(A),$$

where $x(0) = A$, $\dot{x}(0) = y(0) = 0$, and the generalized kinetic [8] and potential energies are

$$T(y) = \int \frac{y\,dy}{1+y^2}, \qquad V(x) = \left(\frac{3}{4}\right)x^{4/3}. \tag{1.3.7}$$

As we will show in the next chapter, all solutions to Eq. (1.3.5) are periodic.

In this volume, we will study TNL oscillators that are members of "odd-parity systems."

Definition 1.3. The differential equation

$$\ddot{x} + F(x, \dot{x}) = 0,$$

is said to be of odd-parity if this equation is invariant under $(x, \dot{x}) \to (-x, -\dot{x})$.

All of the TNL oscillators given in Eq. (1.2.2) are of odd-parity. Another example is given by Eq. (1.3.5).

The real significance of odd-parity systems is that their periodic solutions contain only odd multiples of the fundamental angular frequency in their Fourier series representations [9]. They are also important because many physically relevant systems may be modeled by nonlinear differential equations having this property [2, 7].

Except for Chapter 6, which deals with the possibility of oscillatory, but not necessarily periodic solutions, the remaining chapters will focus on constructing calculational methods for determining analytic expressions for the periodic solutions of TNL oscillators. These TNL oscillatory equations will have the properties of being invariant under time reversal, $t \to -t$, and possessing odd-parity, i.e., invariant under $x \to -x$. As an illustration as to what may occur if both conditions do not simultaneously apply, consider the three equations

$$\ddot{x} + (1 + x\dot{x})x^{1/3} = 0, \tag{1.3.8a}$$

$$\ddot{x} + (1 + \dot{x}^2)x^{1/3} = 0, \tag{1.3.8b}$$

$$\ddot{x} + (1 + \dot{x}^2)x^{1/3} = 0. \tag{1.3.8c}$$

All of these TNL equations are of odd-parity, but the first equation corresponds to a damped oscillator. This can be seen if it is rewritten to the form

$$\ddot{x} + x^{1/3} = -(x x^{1/3})\dot{x}.$$

The right-side represents a damping term and, as a consequence, the solutions are damped and oscillatory rather than periodic. (These results will be shown in Chapter 2.) Note that this equation is of odd-parity, but not invariant under $t \to -t$. The second and third equations are invariant under both $t \to -t$ and $x \to -x$, and all their solutions can be shown to be periodic.

1.4 Scaling and Dimensionless Form of Differential Equations

Differential equations modeling physical phenomena have independent and dependent variables, and parameters appearing in these equations possessing physical units such as mass, length, time, electrical charge, etc. [3, 7, 10]. Thus, the magnitude of these quantities depend on the actual physical units used, i.e., meters versus kilometers, seconds versus hours, etc. A way to eliminate this ambiguity is to reformulate the physical equations such that only dimensionless variables and parameters appear. We now demonstrate how this can be achieved by illustrating the technique on several explicit differential equations. For fuller explanations, see Mickens [3, 7] and de St. Q. Isaacson [10].

1.4.1 *Linear Damped Oscillator*

The modeling differential equation for this physical system is [7, 8]

$$m\frac{d^2x}{dt^2} + k_1\frac{dx}{dt} + kx = 0, \quad x(0) = A, \quad \frac{dx(0)}{dt} = 0, \tag{1.4.1}$$

where m is the mass, k_1 is the damping coefficient, and k is the spring constant. Each term in this equation has the physical units of force and in terms of the units mass (M), length (L) and time (T), we have

$$[x] = L, \quad [t] = T, \quad [\text{force}] = \frac{ML}{T^2}$$

$$[m] = M, \quad [k_1] = \frac{M}{T}, \quad [k] = \frac{M}{T^2}, \quad [A] = L.$$

From the parameters (m, k_1, k) two time scales may be constructed,

$$T_1 = \left(\frac{m}{k}\right)^{1/3}, \qquad T_2 = \frac{m}{k_1}. \tag{1.4.2}$$

With the above indicated initial conditions, there exists only one length scale, i.e.,

$$L_1 = A. \tag{1.4.3}$$

Consequently, the following dimensionless variables can be formed

$$\bar{x} = \frac{x}{L_1} = \frac{x}{A}, \quad \bar{t} = \frac{t}{T_1}. \tag{1.4.4}$$

The particular form for $\bar{t}$ was selected because it is related to the natural frequency of the oscillators in the absence of damping, i.e., $k_1 = 0$; see Fowles [8]. Substitution of $x = A\bar{x}$ and $t = T_1\bar{t}$ into Eq. (1.4.1) gives

$$\begin{aligned}\frac{d^2x}{dt^2} + \left(\frac{k_1}{m}\right)\frac{dx}{dt} + \left(\frac{k}{m}\right)x &= \frac{d^2x}{dt^2} + \left(\frac{1}{T_2}\right)\frac{dx}{dt} + \left(\frac{1}{T_1^2}\right)x \\ &= \left(\frac{A}{T_1^2}\right)\frac{d^2\bar{x}}{d\bar{t}^2} + \left(\frac{A}{T_1T_2}\right)\frac{d\bar{x}}{d\bar{t}} + \left(\frac{A}{T_1^2}\right)\bar{x} = 0,\end{aligned}$$

which upon simplification gives

$$\frac{d^2\bar{x}}{d\bar{t}^2} + \epsilon\frac{d\bar{x}}{dt} + \bar{x} = 0, \quad \bar{x}(0) = 1, \quad \frac{d\bar{x}(0)}{d\bar{t}} = 0, \tag{1.4.5}$$

where

$$\epsilon = \frac{T_1}{T_2}. \tag{1.4.6}$$

Note that the original physical equation contains three parameters (m, k, k_1) and the initial condition parameter A, while the dimensionless equation is expressed in terms of a single parameter ϵ, which can be interpreted as the ratio of the period of the free oscillations to the damping time [7].

1.4.2 *Nonlinear Oscillator*

Consider the Duffing's equation [2, 7]

$$m\frac{d^2x}{dt^2} + kx + k_1x^3 = 0, \quad x(0) = A, \quad \frac{dx(0)}{dt} = 0. \tag{1.4.7}$$

The parameters (m, k, k_1, A) allow the construction of a single time scale

$$T_1 = \left(\frac{m}{k}\right)^{1/2}, \tag{1.4.8}$$

and two length scales

$$L_1 = \left(\frac{k}{k_1}\right)^{1/2}, \quad L_2 = A. \tag{1.4.9}$$

The time scale, T_1, is related to the free oscillations of the linear part of the Duffing equation. The length scale L_1 is an intrinsic, internal scale related only to the *a priori* given properties of the oscillator; it is a consequence of the oscillator being nonlinear. L_2 is the initial condition and thus is an external condition to be imposed on the nonlinear Duffing oscillator.

From the time scale and the two length scales, two dimensionless forms can be obtained; they are

$$x = L_1\bar{x} : \frac{d^2\bar{x}}{d\bar{t}^2} + \bar{x} + \bar{x}^3 = 0, \quad \bar{x}(0) = \frac{A}{L_1}, \quad \frac{d\bar{x}(0)}{d\bar{t}} = 0; \tag{1.4.10}$$

$$x = L_2\bar{x} : \frac{d^2\bar{x}}{d\bar{t}^2} + \bar{x} + \epsilon\bar{x}^3 = 0, \quad \bar{x}(0) = 1, \quad \frac{d\bar{x}(0)}{d\bar{t}} = 0,$$

$$\epsilon = \left(\frac{L_2}{L_1}\right)^2 = \frac{k_1 A^2}{k}. \tag{1.4.11}$$

If $L_2 \ll L_1$, then Eq. (1.4.11) can provide the basis of a standard perturbation approach to solving the Duffing equation. If L_2 and L_2 are of the same order of magnitude, then nonperturbative methods must be applied [7].

1.4.3 $\ddot{x} + ax^p = 0$

Consider the following TNL oscillator

$$\frac{d^2x}{dt^2} + ax^p = 0, \quad x(0) = A, \quad \frac{dx(0)}{dt} = 0, \tag{1.4.12}$$

where

$$a > 0, \quad p = \frac{2m+1}{2n+1}, \quad (m, n) = \text{positive integers}. \tag{1.4.13}$$

Let

$$x = A\bar{x}, \quad t = T\bar{t},$$

and substitute these into Eq. (1.4.12) to obtain

$$\left(\frac{A}{T^2}\right)\frac{d^2\bar{x}}{d\bar{t}^2} + (aA^p)\bar{x}^p = 0,$$

or

$$\frac{d^2\bar{x}}{d\bar{t}^2} + \left[aT^2A^{(p-1)}\right]\bar{x}^p = 0.$$

Setting the coefficient of $\bar{x}^p$ equal to one gives dimensionless equation

$$\frac{d^2\bar{x}}{d\bar{t}^2} + \bar{x}^p = 0, \quad \bar{x}(0) = 1, \quad \frac{d\bar{x}(0)}{d\bar{t}} = 0, \tag{1.4.14}$$

and the time scale

$$T = \left[\frac{A^{(p-1)}}{a}\right]^{1/2}. \tag{1.4.15}$$

1.4.4 $\ddot{x} + ax + bx^{1/3} = 0$

Assume that both a and b are non-negative and consider the TNL oscillator differential equation

$$\frac{d^2x}{dt^2} + ax + bx^{1/3} = 0, \quad \bar{x}(0) = A, \quad \frac{dx(0)}{dt} = 0. \tag{1.4.16}$$

Using

$$t = T\bar{t}, \quad x = L\bar{x},$$

and substituting these expressions into the above differential equation, we find

$$\left(\frac{L}{T^2}\right)\frac{d^2\bar{x}}{d\bar{t}^2} + (aL)\bar{x} + (bL^{1/3})\bar{x}^{1/3} = 0,$$

and

$$\frac{d^2\bar{x}}{d\bar{t}^2} + (aT^2)\bar{x} + \left(\frac{bT^2}{L^{2/3}}\right)\bar{x}^{1/3} = 0.$$

The time and length scales for this particular selection may be calculated by setting to one the coefficients of the second and third terms in the last equation; doing this gives

$$T = \left(\frac{1}{a}\right)^{1/2}, \quad L = \left(\frac{b}{a}\right)^{3/2}. \tag{1.4.17}$$

Inspection of the relation for the length scale indicates that it is an intrinsic value determined by the parameters appearing in the original differential equation, while T_1 is related to the period of the free oscillations when the nonlinear term is absent. In terms of the original initial conditions, given in Eq. (1.4.16), the new initial conditions are

$$\bar{x}(0) = A\left(\frac{a}{b}\right)^{3/2}, \quad \frac{d\bar{x}(0)}{d\bar{t}} = 0. \tag{1.4.18}$$

It is of interest to investigate the case for which the length scale is taken to be $L_1 = A$. For this situation

$$t = T\bar{t}, \quad x = A\bar{x}.$$

Substituting into Eq. (1.4.16) gives

$$\left(\frac{A}{T^2}\right)\frac{d^2\bar{x}}{d\bar{t}^2} + (aA)\bar{x} + (bA^{1/3})\bar{x}^{1/3} = 0$$

and

$$\frac{d^2\bar{x}}{d\bar{t}^2} + (aT^2)\bar{x} + \frac{bT^2}{A^{1/3}}\bar{x}^{1/3} = 0.$$

Setting the coefficient of the $\bar{x}$ term equal to one gives the following expression

$$\frac{d^2\bar{x}}{d\bar{t}^2} + \bar{x} + \epsilon\bar{x}^{1/3} = 0 \tag{1.4.19}$$

where

$$\epsilon = \frac{b}{aA^{2/3}} = \left(\frac{L}{A}\right)^{2/3}. \tag{1.4.20}$$

Observe that ϵ is the ratio of the system's intrinsic length scale to the initial value $x(0) = A$, raised to the two-thirds power. The initial conditions for the dimensionless equation are

$$\bar{x}(0) = 1, \quad \frac{d\bar{x}(0)}{d\bar{t}} = 0.$$

The above example illustrates the fact that we can often eliminate all the parameters and have a nontrivial set of initial conditions or we can have one dimensionless parameter with simple initial conditions. Generally, we will opt for the first situation.

1.5 Exactly Solvable TNL Oscillators

A number of special cases of TNL oscillator differential equations exist that can be solved exactly in terms of standard known functions. These include the antisymmetric, constant force oscillator [11]; the particle-in-a-box [7]; a particular form of the Duffing equation [7]; and the quadratic oscillator [12]. In this section, we present the details of how to obtain the appropriate analytical results for each oscillator.

1.5.1 *Antisymmetric, Constant Force Oscillator* [11]

This oscillator has the following equation of motion

$$\ddot{x} + \text{sgn}(x) = 0, \tag{1.5.1}$$

where the sgn(x) function is

$$\text{sgn}(x) = \begin{cases} +1, & \text{for } x > 0, \\ 0, & \text{for } x = 0, \\ -1, & \text{for } x < 0. \end{cases} \tag{1.5.2}$$

This nonlinear equation is equivalent to the following set of linear equations

$$\ddot{x} + 1 = 0, \quad \text{for } x > 0, \tag{1.5.3a}$$
$$\ddot{x} - 1 = 0, \quad \text{for } x < 0. \tag{1.5.3b}$$

Their respective solutions are

$$x_+(t) = -\left(\frac{1}{2}\right)t^2 + A_1 t + B_1, \tag{1.5.4a}$$
$$x_-(t) = \left(\frac{1}{2}\right)t^2 + A_2 t + B_2, \tag{1.5.4b}$$

where the integration constants are denoted by (A_1, A_2, B_1, B_2). We will now obtain the required solution by using the initial conditions

$$x(0) = 0, \quad \dot{x}(0) = A > 0. \tag{1.5.5}$$

Since $A > 0$, we must use $x_+(t)$ to match these initial conditions, i.e.,

$$x_+(0) = B_1 \Rightarrow B_1 = 0,$$
$$\dot{x}_+(0) = A_1 \Rightarrow A_1 = A,$$

and

$$x_+(t) = -\left(\frac{1}{2}\right)t(t - 2A), \quad 0 \le t \le 2A. \tag{1.5.6}$$

Observe that $x_+(t)$ has the following properties

- $x_+(0) = 0$,
- $x_+(t) > 0$, for $0 < t < 2A$,
- $x_+(2A) = 0$.

These results imply that the period, T, is

$$T = 4A. \tag{1.5.7}$$

Now, at $t = 2A$ and $t = 4A$, we require

$$x_-(2A) = 0, \quad x_-(4A) = 0.$$

From Eq. (1.5.4b) it follows that

$$A_2 = -3A, \quad B_2 = 4A^2,$$

and, as a consequence of these values $x_-(t)$ is

$$x_-(t) = \frac{t^2}{2} - (3A)t + 4A^2, \quad 2A \leq t \leq 4A. \tag{1.5.8}$$

Combining this information gives

$$x(t + 4A) = x(t), \tag{1.5.9}$$

$$x(t) = \begin{cases} -t(t - 2A)/2, & \text{for } 0 \leq t \leq 2A, \\ \dfrac{t^2}{2} - (3A)t + 4A^2, & \text{for } 2A < t \leq 4A. \end{cases} \tag{1.5.10}$$

The Fourier series representation for $x(t)$ can be easily calculated and is given by the following expression

$$x(t) = \left(\frac{16A^2}{\pi^3}\right) \sum_{k=0}^{\infty} \frac{1}{(2k+1)^3} \sin\left[\frac{(2k+1)\pi t}{2A}\right]. \tag{1.5.11}$$

Note that only odd values of the fundamental period, $T = 4A$, appear in the expansion. Further, observe that the four coefficients have the upper bound

$$b_k \leq \frac{C(A)}{k^3}, \tag{1.5.12}$$

where $C(A)$ can be determined by inspection from Eq. (1.5.11).

1.5.2 *Particle-in-a-Box*

Consider a one-dimensional box located between $x = 0$ and $x = L$, i.e., the "size" of the box is L. Let a particle be situated in the box such that at $t = 0$, it is at $x = 0$ with the velocity $v_0 > 0$, i.e., it is moving to the right. After a time T^*, where

$$T^* = \frac{L}{v_0}, \tag{1.5.13}$$

the particle hits the wall at $x = L$, reverses direction and continues to the left. Again, after a time interval of T^*, the particle collides with the wall at $x = 0$ and reverses direction. Thus, the overall motion is periodic with period

$$T = 2T^* = \frac{2L}{v_0}; \tag{1.5.14}$$

therefore

$$x(t+T) = x(t).$$

If we define the velocity function as $v(t) = \dot{x}(t)$, then

$$v(t+T) = v(t),$$

where

$$x(t) = \begin{cases} v_0 t, & \text{for } 0 \le t \le \frac{T}{2}, \\ v_0(T-t), & \text{for } \frac{T}{2} \le t \le T, \end{cases} \tag{1.5.15}$$

$$v(t) = \begin{cases} v_0, & \text{for } 0 < t < \frac{T}{2}, \\ -v_0, & \text{for } \frac{T}{2} < t < T. \end{cases} \tag{1.5.16}$$

If we let $L = \pi$ and $v_0 = 1$, then $T = 2\pi$ and the Fourier series for $x(t)$ and $v(t)$ are

$$x(t) = \frac{\pi}{2} - \left(\frac{4}{\pi}\right) \sum_{k=1}^{\infty} \frac{\cos(2k-1)t}{(2k-1)^2} \tag{1.5.17}$$

$$v(t) = \left(\frac{4}{\pi}\right) \sum_{k=1}^{\infty} \frac{\sin(2k-1)t}{(2k-1)}. \tag{1.5.18}$$

Again, observe that only odd multiples of the fundamental period appear in the expansion.

1.5.3 *Restricted Duffing Equation*

The full Duffing equation takes the form

$$\ddot{x} + k_1\dot{x} + kx + k_2x^3 = 0.$$

The restricted Duffing equation is (in dimensionless units)

$$\ddot{x} + x^3 = 0. \tag{1.5.19}$$

For the initial conditions

$$x(0) - A, \quad \dot{x}(0) = 0, \tag{1.5.20}$$

the exact solution is [7, 13, 14]

$$x(t) = Acn\left(At; 1/\sqrt{2}\right), \tag{1.5.21}$$

where "cn" is the Jacobi elliptic function [13, 15].

Let k and k' satisfy the relation

$$(k')^2 + k^2 = 1.$$

Define the complete elliptical integral of the first kind to be [13, 15]

$$F(k) = \int_0^{\pi/2} \frac{d\theta}{\sqrt{1 - k^2 \sin^2 \theta}}.$$

Define $q(k)$ as

$$q(k) \equiv \exp\left[-\frac{\pi F(k')}{F(k)}\right],$$

and take $v(k, u)$ to be

$$v(k, u) = \left[\frac{\pi}{2F(k)}\right] u.$$

Based on the above quantities, the Jacobi cosine elliptic function is given by the formula

$$cn(u, k) = \left[\frac{2\pi}{kF(k)}\right] \sum_{m=0}^{\infty} \left(\frac{q^{m+\frac{1}{2}}}{1 + q^{2m+1}}\right) \cos(2m + 1)v. \tag{1.5.22}$$

For our case, i.e., the restricted Duffing equation, we have

$$k = \frac{1}{\sqrt{2}}, \quad F\left(\frac{1}{\sqrt{2}}\right) = 1.854\,074\ldots$$

$$q\left(\frac{1}{\sqrt{2}}\right) = 0.043\,213\ldots.$$

If we write $q\left(1/\sqrt{2}\right)$ as

$$q\left(1/\sqrt{2}\right) = e^{-a}$$

then

$$a = 3.141592\ldots$$

and it is easy to show that the Fourier coefficients are bounded by an exponential function of m, i.e.,

$$\begin{aligned} a_m &\equiv \left[\frac{2\sqrt{2}\pi}{F\left(\frac{1}{\sqrt{2}}\right)}\right]\left[\frac{e^{-(m+\frac{1}{2})a}}{1+e^{-(2m+1)a}}\right], \\ &< Ce^{-am}, \end{aligned} \tag{1.5.23}$$

where C can be easily found from the above expression. This result implies that the Fourier coefficients decrease rapidly and, consequently, the use of just a few terms in the expansion of $cn(u,k)$ may provide an accurate analytical representation of the periodic solution [7, 16, 17].

1.5.4 *Quadratic Oscillator*

The quadratic oscillator differential equation is

$$\ddot{x} + |x|x = 0 \quad \text{or} \quad \ddot{x} + x^2\text{sgn}(x) = 0. \tag{1.5.24}$$

In Section 2.2.2, we show that all solutions for this TNL oscillator are periodic with period given by the expression [12]

$$T(A) = \frac{2^{1/6}\left[\Gamma\left(\frac{1}{3}\right)\right]^3}{\pi} \cdot \frac{1}{A^{1/2}}, \tag{1.5.25}$$

where $x(0) = A$ and $\dot{x}(0) = 0$ are the initial conditions. Further, we find that the solution is

$$x(t) = A\left[\frac{(\sqrt{3}+1)\,cn(t,k) - (\sqrt{3}-1)}{1 + cn(t,k)}\right], \tag{1.5.26}$$

where

$$cn(t,k) \equiv cn\left[\left(\frac{4A^2}{3}\right)^{1/4} t, k\right], \quad k^2 = \frac{2+\sqrt{3}}{4}. \tag{1.5.27}$$

Consequently, the periodic solution is expressed as a rational function of the Jacobi cosine function.

1.6 Overview of TNL Oscillator Methods

Nonlinear oscillations occurring in one-degree-of-freedom systems have been studied intensely for almost two centuries [2–4, 18–25]. The general form that those equations take is

$$\ddot{x} + x = \epsilon f(x, \dot{x}), \quad 0 < \epsilon \ll 1, \tag{1.6.1}$$

where ϵ is a small parameter. These classical methods are based on expansions in terms of ϵ which are taken to be asymptotic series. Each particular perturbation method is distinguished by how this feature is accomplished. If from *a priori* considerations it can be determined that periodic solutions exist, then a major task, for each method, is to eliminate the so-called secular terms. Secular terms are expressions in the solutions that are oscillatory, with increasing, time dependent amplitudes [3, 4, 7], i.e., for an odd-parity system

$$\text{secular term} : t^n \cos[(2k+1)\Omega t], \tag{1.6.2}$$

where (n, k) are integers, with $n \geq 1$ and $k \geq 0$. For all of the standard methods, procedures have evolved to resolve this issue.

Inspection of Eq. (1.6.1) shows that each of the classical methods has at its foundation the explicit assumption that when $\epsilon = 0$ the resulting "core" equation is the linear harmonic oscillator differential equation, namely,

$$\ddot{x}_0 + x_0 = 0, \tag{1.6.3}$$

where the zero indicates $\epsilon = 0$. This fact presents an immediate difficulty for TNL oscillators, where Eq. (1.6.1) is replaced by, for example,

$$\ddot{x} + x^p = \epsilon f(x, \dot{x}), \quad p \neq 1. \tag{1.6.4}$$

We observe that when $\epsilon = 0$, this equation reduces to the nonlinear equation

$$\ddot{x}_0 + x_0^p = 0, \tag{1.6.5}$$

and this type of equation would, at the very least, greatly complicate any solution construction based on expansions in the parameter ϵ. The general conclusion is that the standard classical perturbation procedures cannot be applied to TNL equations.

Generally, four techniques can be used to determine the approximations to the periodic solutions of nonlinear oscillator differential equations. In general, they may be applied to both standard and TNL equations, i.e., those that can be expressed as in Eqs. (1.6.1) and (1.6.4). All of these procedures, except for one, set up a methodology that converts the problem of solving a single, second-order, nonlinear differential equation to one of solving, in sequence, an infinite set of linear, inhomogeneous equations. A brief discussion of each procedure will now be given. The relevant details will appear in the chapter devoted to each particular method.

1.6.1 *Harmonic Balance*

The harmonic balance method is based on the use of an assumed truncated trigonometric expansion for the periodic solution. The n-th order approximation takes a form such as

$$x_n(t) = a_1 \cos\theta + a_2 \cos(3\theta) + \cdots + a_n \cos(2n-1)\theta, \tag{1.6.6}$$

where $\theta = \Omega_n t$, and the n-coefficients and Ω_n are to be determined. For a conservative system with initial conditions, $x(0) = A$ and $\dot{x}(0) = 0$, the basic strategy is to substitute Eq. (1.6.6) into the differential equation and expand the resulting expression into a trigonometric series, but only including terms from $\cos\theta$ to $\cos(2n-1)\theta$; doing this gives the following type of relation

$$\begin{aligned} &H_1(a_1, a_2, \ldots, a_n, \Omega_n)\cos\theta + H_2(a_1, a_2, \ldots, a_n, \Omega_n)\cos(3\theta) \\ &\quad + \cdots + H_n(a_1, a_2, \ldots, \Omega_n)\cos(2n-1)\theta + \text{HOH} \simeq 0, \end{aligned} \tag{1.6.7}$$

where HOH stands for higher-order-harmonics, and for a given differential equation the $H_i(a_1, a_2, \ldots, a_n, \Omega_n)$ are completely specified. The harmonic balancing procedure consists in setting the coefficients of the cosine terms to zero, i.e.,

$$H_i(a_1, a_2, \ldots, a_n, \Omega_n) = 0; \quad i = 1, 2, 3, \ldots, n. \tag{1.6.8}$$

These n-relations, along with the initial conditions, can be solved to give all the coefficients and Ω_n as functions of A, i.e.,

$$\begin{cases} a_i = a_i(A); & i = 1, 2, \ldots, n; \\ \Omega_n = \Omega_n(A). \end{cases} \tag{1.6.9}$$

1.6.2 *Parameter Expansion*

The parameter expansion method is an extension and generalization of standard perturbation methods. The basic idea is to take a parameter occurring in the differential equation and represent it as an expansion in terms of a fictitious or artificial parameter. Such an expansion may provide a valid (asymptotic) solution when the parameter has small values. If this is correct, the resulting expression is then evaluated for some large value of the parameter (usually selected to be one) for the actual problem of interest.

To illustrate the method, consider the TNL equation

$$\ddot{x} + x^3 = 0, \tag{1.6.10}$$

it can be rewritten as

$$\ddot{x} + \Omega^2 x = \Omega^2 x - x^3, \tag{1.6.11}$$

where Ω^2 is, for the present, unknown. Now a parameter p is introduced in the following way

$$\ddot{x} + \Omega^2 x = p(\Omega^2 x - x^3). \tag{1.6.12}$$

The basic idea is now to assume p to be small and treat the latter equation using standard perturbation methods. After this is done, one sets $p = 1$ in the resulting expressions for the approximations to the solutions. In general, the angular frequency, $\Omega = 2\pi/T$, is determined by the requirement that no secular terms be present in the solutions.

1.6.3 *Averaging Methods*

Averaging methods generally start with the following representation for the periodic solutions

$$x(t) = a(t)\cos\psi(t), \tag{1.6.13}$$

and, through a series of assumptions and mathematical manipulations, derive first-order differential equations for unknown functions $a(t)$ and $\psi(t)$, i.e.,

$$\frac{da}{dt} = \epsilon F_1(a), \quad \frac{d\psi}{dt} = \epsilon F_2(a), \tag{1.6.14}$$

where ϵ is, in general, some small parameter appearing in the original differential equation. These equations are solved, in the order presented, to obtain $a(t,\epsilon)$ and $\psi(t,\epsilon)$.

In somewhat more detail, the "a" and "ψ" in Eq. (1.6.14) are not, strictly speaking the same as the "a" and "ψ" in Eq. (1.6.13). The latter equations are averaged equations, with the averaging done on an interval of 2π in the variable ψ. The details as to what exactly is needed to obtain the expressions in Eq. (1.6.14) will be given later in the chapter devoted to this topic.

This method has the distinct advantage, as compared with all the other procedures, of allowing oscillatory, but not necessarily periodic solutions, to be calculated.

1.6.4 *Iteration Techniques*

Iteration techniques start with a given nonlinear (regular or TNL) oscillator differential equation and, through a series of manipulations, transform it into a set of linear, second-order, inhomogeneous equations that must be solved sequentially. For example, consider

$$\ddot{x} + x^3 = 0, \quad x(0) = A, \quad \dot{x}(0) = 0,$$

and add $\Omega^2 x$ to both sides to obtain

$$\ddot{x} + \Omega^2 x = \Omega^2 x - x^3. \tag{1.6.15}$$

Define $x_0(t)$ to be

$$x_0(t) = A\cos(\Omega_0 t),$$

where Ω_0 is currently known. Now define the sequence of functions

$$x_0(t), x_1(t), \ldots, x_n(t), \ldots,$$

which are solutions to

$$\begin{cases} \ddot{x}_{k+1} + \Omega_k^2 x_{k+1} = \Omega_k^2 x_k - x_k^3, \\ x_{k+1}(0) = A, \quad \dot{x}_{k+1}(0) = 0. \end{cases} \tag{1.6.16}$$

The Ω_k are determined by the requirement that x_{k+1} does not contain secular terms. Note that under these conditions, $x_1(t)$ can be calculated from a knowledge of $x_0(t)$; $x_2(t)$, likewise, can be determined from $x_1(t)$; etc. In practice, the hope is that $x_k(t)$, for small values of k, will provide an accurate representation to the actual periodic solutions.

1.7 Discussion

We end this chapter with a brief overview of some of the criteria or qualities expected of a method that can be used to calculate analytical approximations to the periodic solutions to a TNL oscillator. However, in the final analysis, the validity and value of a particular method and the solutions that it produces depend heavily on what we intend to do with the results obtained from the calculations. However, the following four items/issues are of prime importance:

1) The method of calculation should be rather direct to understand and easy to implement.

In particular, this means that higher-order approximations should be capable of being straightforwardly obtained, although in practice it may be algebraically intensive to carry out this process.

2) The method should allow accurate estimates to be made for the period of the oscillations.

Again, what is to be considered "accurate enough" will clearly be a function of what we intend to do with this result.

3) The calculational procedure should produce trigonometric approximations to the actual Fourier expansions such that *a priori* known restrictions on the expansion coefficients (bounds, rates of decrease, etc.) should be generally satisfied by those of the approximate solutions.

A wide range of theorems exist on the general properties of Fourier coefficients [16, 17]. To the degree that this is possible, the coefficients appearing in the approximate expressions should also satisfy these restrictions. In fact, these limitations on the coefficients may be used as a measure of the "quality" of the solutions produced by a given calculational scheme.

4) The approximate solutions obtained from a particular calculational scheme should have the appropriate mathematical forms, structures, and properties known to be possessed by the exact solutions.

Thus, for example, the TNL equation is of odd-parity, then the approximate solutions should only contain odd multiples of the fundamental period. The occurrence of terms having even periods would indicate that the scheme is incorrect. In a similar fashion, if the TNL oscillator is conservative and if the initial conditions, $x(0) = A$ and $\dot{x}(0) = 0$, are selected, then only cosine terms should appear in its trigonometric approximation expansion.

While we realize that the criteria presented above are vague and heuristic in nature, it is clear that this situation is as it is because approximation procedures in practice, and in the results they produce, are by their essence never fully based on rigorous mathematics. In almost all cases, some mathematical requirement that was used to justify the procedure is violated. Therefore, the real value of a (practical) approximation method is whether it provides a suitable resolution of some set of issues related to a problem that is formulated in the language of mathematics. This means that the user of such schemes must have deep fundamental insight into the original (in most cases) physical problem, while also understanding and acknowledging the limitations of a purely mathematical approach. This realization has provided much of the existing tension between "pure" and "applied" mathematics.

Problems

1.1 Which of the following functions are TNL (at $x = 0$)?
  (i) x^2
  (ii) $x^{1/2} + x$
  (iii) $x + \frac{\text{sgn}(x)}{|x|^{1/2}}$
  (iv) $x + x^{5/3}$
  (v) $x + |x|$.

1.2 Are any of the following differential equations TNL equations? Why?
  (i) $\ddot{x} + \frac{1}{x} = 0$
  (ii) $\ddot{x} - x + x^3 = 0$
  (iii) $\ddot{x} + x + \frac{1}{x^{3/5}} = 0$
  (iv) $\ddot{x} + \frac{x^{5/3}}{1+x^2} = 0$.

1.3 Prove that

$$\frac{d}{dx}|x| = \text{sgn}(x).$$

1.4 Is x^3 the same as $|x|^2 x$. Explain your answer. What about x and $|x|\text{sgn}(x)$?

1.5 Transform the following differential equations to dimensionless forms.
  (i) $m\ddot{x} + ax^2 + bx^3 = 0$
  (ii) $\ddot{x} + \omega^2 x + kx^{1/3} = (a - bx^2)\dot{x}$
  (iii) $\ddot{x} + \frac{\lambda x^3}{1+fx^2} = 0$
  (iv) $\ddot{x} + g|x|^2\text{sgn}(x) + hx^3 = 0$.

  Assume the initial conditions are $x(0) = A$ and $\dot{x}(0) = 0$.

  - For a given equation, are the scales unique?
  - If not, discuss the differences between the scales.
  - What physical interpretation can be associated with each set of scales when a particular differential equation has more than one set of scales.

1.6 Are any of the differential equations listed in Problems 1.2 and 1.5 of odd-parity? Which are invariant under $t \to -t$?

References

[1] B. van der Pol, *Philosophical Magazine* (1926) 978; **3** (1927) 65.
[2] A. H. Nayfeh and D. T. Mook, *Nonlinear Oscillations* (Wiley-Interscience, New York, 1979).
[3] R. E. Mickens, *Nonlinear Oscillations* (Cambridge University Press, New York, 1981).
[4] J. A. Murdock, *Perturbations: Theory and Methods* (Wiley-Interscience, New York, 1981).
[5] R. E. Mickens and K. Oyedeji, *Journal of Sound and Vibration* **102** (1985) 579.
[6] R. E. Mickens, *Journal of Sound and Vibration* **292** (2006) 964.
[7] R. E. Mickens, *Oscillations in Planar Dynamic Systems* (World Scientific, Singapore, 1996).
[8] G. R. Fowles, *Analytical Mechanics* (Holt, Rinehart, and Winston; New York, 1962).
[9] R. E. Mickens, *Journal of Sound and Vibration* **258** (2002) 398.
[10] E. de St. Q. Isaacson and M. de St. Q. Isaacson, *Dimensional Methods in Engineering and Physics* (Wiley, New York, 1975).
[11] T. Lipscomb and R. E. Mickens, *Journal of Sound and Vibration* **169** (1994) 138.
[12] R. E. Mickens, unpublished results, May 2009.
[13] P. F. Boyd and M. D. Friedman, *Handbook of Elliptic Integrals for Engineers and Physicists* (Springer-Verlag, Berlin, 1954).
[14] H. T. Davis, *Introduction to Nonlinear Differential and Integral Equations* (Dover, New York, 1962).
[15] R. E. Mickens, *Mathematical Methods for the Natural and Engineering Sciences* (World Scientific, London, 2004). See Section 3.9.
[16] D. C. Champeny, *A Handbook of Fourier Theorems* (Cambridge University Press, Cambridge, 1987).
[17] T. W. Körner, *Fourier Analysis* (Cambridge University Press, Cambridge, 1988).
[18] H. Poincaré, *New Methods in Celestial Mechanics*, Vols. I–III (English translation), NASA TTF-450, 1967.
[19] A. A. Andronov and C. E. Chaikin, *Theory of Oscillations* (Princeton University Press, Princeton, NJ; 1949).
[20] N. N. Bogoliubov and Y. A. Mitropolsky, *Asymptotic Methods in the Theory of Non-linear Oscillations* (Hindustan Publishing, Delhi, 1961).
[21] J. J. Stoker, *Nonlinear Vibrations* (Wiley-Interscience, New, York, 1950).
[22] N. Minorsky, *Nonlinear Oscillations* (Van Nostrand Reinhold; Princeton, NJ; 1962).
[23] J. K. Hale, *Oscillations in Nonlinear Systems* (McGraw-Hill, New York, 1963).
[24] R. H. Rand and D. Armbruster, *Perturbation Methods Bifurcation Theory and Computer Algebra* (Springer-Verlag, New York, 1987).

[25] A. W. Bush, *Perturbation Methods for Engineers and Scientists* (CRC Press; Boca Raton, FL; 1992).

Chapter 2

Establishing Periodicity

Before any attempt is made to calculate approximations to the periodic solutions of either standard or TNL oscillator differential equations, we must provide justifications for why we believe periodic solutions exist for the particular equation of interest. We will always assume that our differential equations satisfy the appropriate conditions such that an existence and uniqueness theorem holds for solutions [1–4].

This chapter examines and applies some general techniques that can be used to illustrate the existence of periodic solutions for a given TNL equation. These methods also apply to the case of standard equations [5]. The first section introduces the notion of a two-dimension (2-dim) phase-space and explains how the ideas associated with this concept may be used to determine whether a given equation has periodic solutions. In Section 2.2, we apply the results of Section 2.1 to a number of TNL oscillatory differential equations. Since many systems are influenced by friction or dissipative forces, Section 2.3 provides a procedure for obtaining useful information for this case. Finally, in the last section, we give a concise summary of what was achieved in this chapter.

2.1 Phase-Space [5]

All of the nonlinear, second-order, differential equations examined in this volume can be written as special cases of the equation

$$\ddot{x} + f(x) = \epsilon g(x, \dot{x}),$$

where ϵ is a parameter, and as explained previously, the dot notation indicates differentiation with respect to time, i.e., $\dot{x} \equiv dx/dt$ and $\ddot{x} \equiv d^2x/dt^2$.

However, for this section, we only consider the equation

$$\ddot{x} + f(x) = 0. \tag{2.1.1}$$

We now examine Eq. (2.1.1) within the framework of a 2-dim phase-space.

2.1.1 *System Equations*

The second-order differential equation, given in Eq. (2.1.1), may be reformulated to two first-order system equations

$$\dot{x} = y, \quad \dot{y} = -f(x). \tag{2.1.2}$$

The first equation is a definition of the variable y, while the second equation contains all the dynamics of the original second-order differential equation. Observe that this method of constructing the system equations is not unique; another valid representation is

$$\dot{x} = -y, \quad \dot{y} = f(x).$$

The variables x and y define a 2-dim phase-space which we denote as (x, y).

2.1.2 *Fixed-Points*

The fixed-points are constant or equilibrium solutions to the system equations, i.e., they correspond to $x(t) =$ constant and $y(t) =$ constant. Therefore the fixed-points are simultaneous solutions to

$$\bar{y} = 0, \quad f(\bar{x}) = 0, \tag{2.1.3}$$

where the barred quantities indicate the constant solutions, i.e.,

$$x(t) = \bar{x}, \quad y(t) = \bar{y}.$$

For systems modeled by Eq. (2.1.1), the fixed-points are all located on the phase-space x-axis. If $f(\bar{x}) = 0$ has m-real solutions $(\bar{x}_1, \bar{x}_2, \ldots, \bar{x}_m)$, then the fixed-points are

$$(\bar{x}_1, 0), (\bar{x}_2, 0), \ldots, (\bar{x}_m, 0).$$

2.1.3 *ODE for Phase-Space Trajectories*

The solutions to Eq. (2.1.2) trace out curves in the (x, y) phase-space, i.e., $(x(t), y(t))$ as the solutions evolve with time, t. These are the phase-space trajectories and they satisfy a differential equation that we will now derive.

Let $x(t)$ and $y(t)$ be solutions to Eq. (2.1.2), subject to initial conditions $x(0) = x_0$ and $y(0) = y_0$, where (x_0, y_0) is given. Then, as stated above, for all $t > 0$, the point $(x(t), y(t))$ moves in phase-space producing a curve $\mathcal{C}(x, y)$. The differential equation satisfied by this curve can be determined by the following argument. Let $y = y(x)$, be the equation of the curve; then

$$\frac{dy}{dt} = \frac{dy}{dx}\frac{dx}{dt},$$

and using the results in Eq. (2.1.2), we obtain

$$\frac{dy}{dx} = -\frac{f(x)}{y}. \tag{2.1.4}$$

In general, this is a first-order, nonlinear differential equation, whose solutions are the curves of the solution trajectories in phase-space.

2.1.4 *Null-clines*

Null-clines are curves in phase-space along which the derivative, $y' = dy/dx$, has constant values. Null-clines, in general, are not solutions to the trajectory differential equation; however, they help organize phase-space in a manner to be discussed below.

Our interest is in only two particular null-clines: the curves along which $y' = 0$ and $y' = \infty$. We denote these two curves, respectively by $y_0(x)$ and $y_\infty(x)$.

Examination of Eq. (2.1.4) gives the following results for $y_0(x)$ and $y_\infty(x)$:

$$\begin{cases} y' = 0: \text{ Along the curves } \bar{x} = 0, \text{ where } \bar{x} \text{ is a real solution of } f(\bar{x}) = 0. \\ y' = \infty: \text{ Along the } x\text{-axis.} \end{cases}$$

Thus, $y_0(x)$ consists of vertical lines, $\bar{x}$ = constant, corresponding to the real solutions of $f(\bar{x}) = 0$; while $y_\infty(x)$ is the full x-axis.

This analysis reveals that the $y_0(x)$ and $y_\infty(x)$ null-clines always intersect at a fixed point.

A second feature, of equal significance is that the two null-clines divide the phase-plane into a number of open regions. In each of these regions, the sign of the derivative, dy/dx, is fixed, i.e., it is either positive or negative, and, furthermore, dy/dx is bounded. Therefore, the only places where the derivative can be zero or unbounded is on one of the two null-clines and they form the boundaries of the open regions, for which only a single sign for dy/dx can occur [5, 6].

2.1.5 *Symmetry Transformations*

A symmetry transformation is a change of dependent variables such that the form of the original differential equation is maintained by the transformed equation expressed in terms of the new variables.

Let the system equations (in a more general form than we have considered) be written as

$$\dot{x} = F(x, y), \quad \dot{y} = G(x, y), \tag{2.1.5}$$

and make the change of variables

$$x = T_1(\tilde{x}, \tilde{y}), \quad y = T_2(\tilde{x}, \tilde{y}), \tag{2.1.6}$$

where $\tilde{x}$ and $\tilde{y}$ are the new dependent variables. If on substitution of Eq. (2.1.6) into Eq. (2.1.5), the new system equations take the form

$$\dot{\tilde{x}} = F(\tilde{x}, \tilde{y}), \quad \dot{\tilde{y}} = G(\tilde{x}, \tilde{y}), \tag{2.1.7}$$

then Eq. (2.1.6) is said to be a symmetry transformation of Eq. (2.1.5). This means that (x, y) and $(\tilde{x}, \tilde{y})$ satisfy exactly the same differential equations.

2.1.6 *Closed Phase-Space Trajectories* [3, 5, 7]

Simple, closed curves in phase-space correspond to periodic solution. This follows from the fact that on completing a path from the original state, $x(t_0)$ and $y(t_0)$, to that same state at time $t_0 + T$, the existence and uniqueness theorems require that the motion continues to repeat this same behavior indefinitely in time. Since

$$x(t_0 + T) = x(t_0), \quad y(t_0 + T) = y(t_0),$$

then the period of this motion is T [7].

2.1.7 *First-Integrals*

The first-order differential equation for the path of the trajectories in phase-space is separable; see, Eq. (2.1.4),

$$\frac{dy}{dx} = -\frac{f(x)}{y},$$

and can be immediately integrated to obtain the relation

$$\frac{y^2}{y} + V(x) = \text{constant}, \tag{2.1.8}$$

where the potential function, $V(x)$, is defined to be

$$V(x) = \int f(x)dx. \tag{2.1.9}$$

This equation is called a first-integral of the differential equation

$$\ddot{x} + f(x) = 0.$$

In the next section, we use the various concepts introduced in this section to demonstrate the existence of periodic solutions for a wide range of TNL oscillator equations.

2.2 Application of Phase-Space Methods

2.2.1 *Linear Harmonic Oscillator*

The linear harmonic oscillator provides a model for a broad range of phenomena in the natural and engineering sciences [8–10]. The differential equation for this system is

$$\ddot{x} + x = 0, \quad x(0) = A, \quad \dot{x}(0) = 0, \tag{2.2.1}$$

and its exact solution is

$$x(t) = A\cos t. \tag{2.2.2}$$

We now follow the procedures of Section 2.1 and demonstrate independently that the linear harmonic oscillator equation has all periodic solutions:

(i) The *system equations* are

$$\frac{dx}{dt} = y, \quad \frac{dy}{dt} = -x. \tag{2.2.3}$$

(ii) A single *fixed-point* exists and it is located at

$$(\bar{x}, \bar{y}) = (0, 0). \tag{2.2.4}$$

(iii) The *trajectory equation* is

$$\frac{dy}{dx} = -\frac{x}{y}. \tag{2.2.5}$$

(iv) The two *null-clines* are

$$\begin{cases} y' = 0: & \text{along the } y\text{-axis,} \\ y' = \infty: & \text{along the } x\text{-axis.} \end{cases} \tag{2.2.6}$$

(v) The trajectory equation Eq. (2.2.5) is *invariant* under the three coordinate transformations

$$S_1 : x \to -x, \quad y \to y \tag{2.2.7a}$$
(reflection in the y-axis)

$$S_2 : x \to x, \quad y \to -y \tag{2.2.7b}$$
(reflection in the x-axis)

$$S_3 : x \to -x, \quad y \to -y \tag{2.2.7c}$$
(reflection/inversion through the origin).

We will call these transformations the *symmetries* of the original differential equation (2.2.1).

(vi) A first-integral exists and it is

$$y^2 + x^2 = A^2. \tag{2.2.8}$$

With these results, we now prove that all the solutions to the linear harmonic oscillator are periodic. This can be achieved in two ways.

First, observe that the trajectories in phase-space are given by Eq. (2.2.8), the first-integral. The corresponding curve is closed (in fact, a circle) for any value of $A \neq 0$, and therefore the conclusion is that all solutions must be periodic. Note that $A = 0$ gives the fixed-point $(\bar{x}, \bar{y}) = (0, 0)$.

A second method for demonstrating that all solutions are periodic is to use the known geometrical properties of the associated phase-plane. Figure 2.2.1 gives the essential features. In particular:

- The null-clines, $y_0(x)$ and $y_\infty(x)$, lie along the respectively y and x axes.
- The null-clines, $y_0(x)$ and $y_\infty(x)$, divide the phase-plane into four open regions, each coinciding with a quadrant of the plane.
- In each quadrant, the sign of the derivative, dy/dx, has a definite value.

The proof that all trajectories are closed proceeds as follows (see Figure 2.2.2):

1) Select an arbitrary point, P_1, on the y-axis. The trajectory through this point has to have zero slope at P_1, decrease in the first quadrant, and

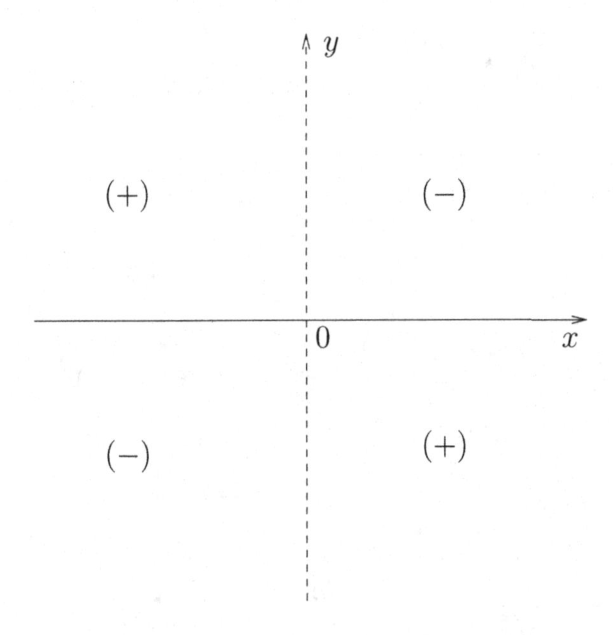

Fig. 2.2.1 Basic properties of the phase-plane for the linear harmonic oscillator. The dashed-line (- - -) is the "zero" null-cline, the solid line (—) is the "infinite" null-cline. The ($\pm$) indicates the sign of dy/dx for the designated region.

intersect the x-axis with unbounded (infinite) slope. See part (b) of the diagram.

2) The application of the symmetry, S_2, i.e., reflection in the x-axis, gives the result in (c).

3) Finally, the application of the symmetry, S_1, i.e., reflection in the y-axis, gives (d).

4) Since reflection symmetries produce images that are continuous at the line through which the reflection is made, we may conclude that the net result of all our operations is to generate a closed curve in the phase-plane. Consequently, this path corresponds to a periodic solution.

5) The point P_1 is an arbitrary selection. (Note that it is only for our convenience that P_1 was selected to lie on the y-axis; any other choice would work, but the total effort to show that the trajectory through this point is a closed curve would be greater.) Therefore, we conclude that all solutions are periodic.

General Comments: At no point in the above phase-plane arguments did the actual form of the differential equation need to be known. This implies that for any particular differential equation, independently as to whether

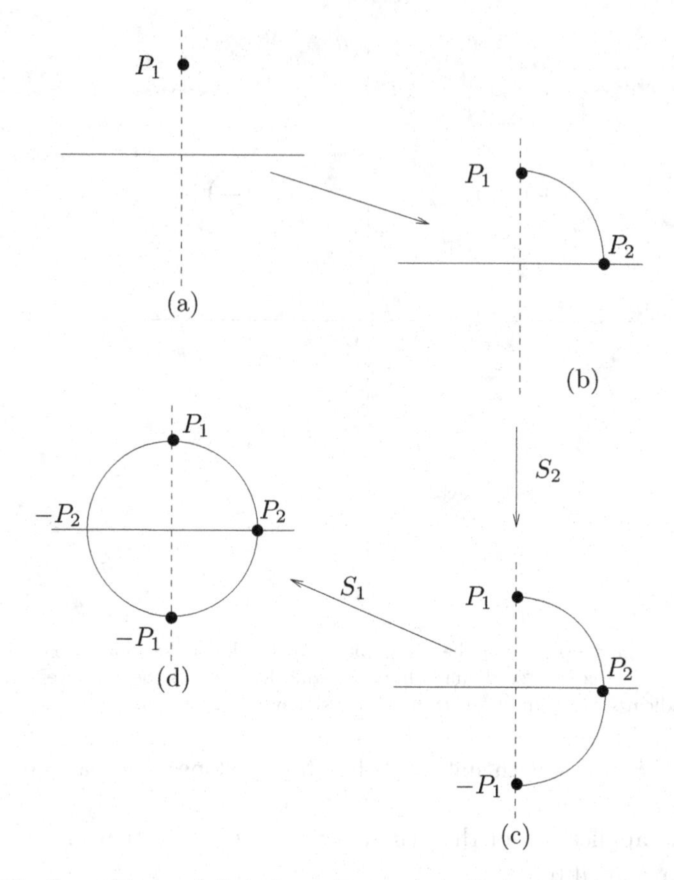

Fig. 2.2.2 Geometric proof that phase-plane trajectories are closed curves using the symmetry transformations.

it is a regular or TNL oscillator, if it has the mathematical structure such that the following phase-plane properties exist, then all the solutions are periodic. The required features are:

(i) There is a single fixed-point at $(\bar{x}, \bar{y}) = (0, 0)$.
(ii) The $y_0(x)$ and $y_\infty(x)$ null-clines coincide, respectively, with the y and x axes.
(iii) The four open domains to which the phase-plane are divided by the null-clines are as presented in Figure 2.2.1.
(iv) The trajectory equation, for the phase-plane curves, is invariant under the transformations S_1, S_2, and S_3.

This is an important result and will be used in the next subsection to prove that several TNL oscillator equations have only periodic solutions.

2.2.2 *Several TNL Oscillator Equations*

The following six second-order, nonlinear differential equations are examples of TNL oscillators:

$$\begin{aligned} &\ddot{x} + x^3 = 0, \\ &\ddot{x} + x^{3/5} = 0, \\ &\ddot{x} + x + x^{1/3} = 0, \\ &\ddot{x} + x^2 \text{sgn}(x) = 0, \\ &\ddot{x} + (1 + \dot{x}^2) x^{1/3} = 0, \\ &\ddot{x} + \frac{1}{x^{1/3}} = 0. \end{aligned}$$

Close inspection of all these equations shows that they possess the following properties:

(a) They are invariant under time inversion, $t \to -t$, and are of odd-parity.
(b) They all have a single fixed-point, located in the phase-plane at $(\bar{x}, \bar{y}) = (0, 0)$.
(c) Each has null-clines, $y_0(x)$ and $y_\infty(x)$, that coincide, respectively, with the y and x axes.
(d) Each has a trajectory equation that is invariant under S_1, S_2, and S_3.
(e) Their respective phase-planes may be represented as given in Figure 2.2.1.
(f) They all have first-integrals that can be explicitly calculated.

From the totality of properties, given in a) to f), we conclude, based on the results of Section 2.2.1, that all of the above listed TNL oscillators have only periodic solutions.

We now examine the fourth equation listed above, i.e.,

$$\ddot{x} + x^2 \text{sgn}(x) = 0, \tag{2.2.9}$$

and calculate its exact solution. To begin, take the initial conditions to be

$$x(0) = A, \quad \dot{x}(0) = y(0) = 0. \tag{2.2.10}$$

The trajectory equation and first-integral are given, respectively, by the relations

$$\frac{dy}{dx} = -\frac{x^2 \text{sgn}(x)}{y}, \tag{2.2.11}$$

$$\frac{y^2}{2} + \left(\frac{x^3}{3}\right)\operatorname{sgn}(x) = \frac{A^3}{3}. \tag{2.2.12}$$

Since $y = dx/dt$, then in the fourth-quadrant of the phase-plane, where y is negative and x positive, we have

$$y = \frac{dx}{dt} = -\left(\sqrt{\frac{2}{3}}\right)\sqrt{A^3 - x^3},$$

or

$$dt = -\left(\sqrt{\frac{3}{2}}\right)\frac{dx}{\sqrt{A^3 - x^3}}.$$

Because of the symmetry properties, the period of the oscillation can be calculated from the expression

$$\int_0^{T/4} dt = -\left(\sqrt{\frac{3}{2}}\right)\int_A^0 \frac{dx}{\sqrt{A^3 - x^3}}, \tag{2.2.13}$$

and this can be written as

$$T(A) = 4\left(\sqrt{\frac{3}{2}}\right)\int_A^0 \frac{dx}{\sqrt{A^3 - x^3}}. \tag{2.2.14}$$

Let $x = Az$, then

$$T(A) = 4\left(\sqrt{\frac{3}{2}}\right)\left(\frac{1}{A^{1/2}}\right)\int_0^1 \frac{dz}{\sqrt{1 - z^3}}. \tag{2.2.15}$$

From [12–14], we have

$$\int_0^1 \frac{dz}{\sqrt{1 - z^3}} = \left[\frac{1}{2\pi\sqrt{3}(2^{1/3})}\right]\left[\Gamma\left(\frac{1}{3}\right)\right]^3, \tag{2.2.16}$$

where $\Gamma\left(\frac{1}{3}\right) = 2.678\,938\,534\ldots$. Therefore, the period, $T(A)$, is

$$T(A) = \frac{\left[\Gamma\left(\frac{1}{3}\right)\right]^3 (2^{1/6})}{\pi A^{1/2}}, \tag{2.2.17}$$

with angular frequency equal to

$$\Omega(A) = \frac{2\pi}{T(A)} = \left\{\frac{\pi^2(2^{5/6})}{\left[\Gamma\left(\frac{1}{3}\right)\right]^3}\right\} A^{1/2}. \tag{2.2.18}$$

In a similar manner, the exact solution can be calculated. Starting with

$$-\frac{dx}{\sqrt{A^3 - x^3\operatorname{sgn}(x)}} = \sqrt{\frac{2}{3}}\,dt,$$

it follows that

$$-\int_A^x \frac{du}{\sqrt{A^3 - u^3 \text{sgn}(u)}} = \sqrt{\frac{2}{3}}\, t. \tag{2.2.19}$$

Now let $u = Az$, then the left-hand side of Eq. (2.2.19) becomes

$$\int_x^A \frac{du}{\sqrt{A^3 - u^3}} = \left(\frac{1}{A^{1/2}}\right) \int_{x/A}^1 \frac{dz}{\sqrt{1 - z^3}}. \tag{2.2.20}$$

But [13, 14],

$$\int_v^1 \frac{dz}{\sqrt{1 - z^3}} = \left(\frac{1}{3^{1/4}}\right) cn^{-1}\left(\frac{\sqrt{3} - 1 + v}{\sqrt{3} + 1 - v}, k\right), \tag{2.2.21}$$

where cn^{-1} is the inverse Jacobi cosine function and

$$k^2 = \frac{2 + \sqrt{3}}{4}. \tag{2.2.22}$$

Therefore, from Eqs. (2.2.19), (2.2.20) and (2.2.21) it follows that for $v = x/A$,

$$\frac{\sqrt{3} - 1 + v}{\sqrt{3} + 1 - v} = cn\left[\left(\frac{4A^2}{3}\right)^{1/4} t, k\right].$$

Solving for $x = Av$ gives

$$x(t) = A\left[\frac{(\sqrt{3} + 1)\, cn(t, k) - (\sqrt{3} - 1)}{1 + cn(t, k)}\right], \tag{2.2.23}$$

where we have used a shorthand notation for the Jacobi cosine function. Equation (2.2.23) is the solution for the TNL oscillator given by Eq. (2.2.9).

2.3 Dissipative Systems: Energy Methods [7]

For completeness, we now discuss conservative oscillators containing a positive damping term. For our purposes, the equation of motion takes the form

$$\ddot{x} + f(x) = -\epsilon g(x)\dot{x}, \quad x(0) = A, \quad \dot{x}(0) = 0, \tag{2.3.1}$$

where the right-hand side is the damping term and it is assumed that

$$g(x) \geq 0, \quad -\infty < x < \infty; \quad \epsilon > 0. \tag{2.3.2}$$

Other more general forms exist for damping terms, however, their use does not lead to any fundamental changes in the conclusions to be reached.

For $\epsilon = 0$, we have

$$\ddot{x} + f(x) = 0,$$

and this equation has the first-integral

$$\frac{y^2}{2} + V(x) = V(A), \quad V(x) = \int f(x)dx.$$

Let $V(x)$, the potential, have the properties

$$V(0) = 0, \quad V(-x) = V(x); \quad V(x) > 0, \quad x \neq 0.$$

If this is true, then the function

$$W(x, y) \equiv \frac{y^2}{2} + V(x) \tag{2.3.3}$$

satisfies the condition

$$W(x, y) > 0 \quad \text{if } x \neq 0, \quad y \neq 0; \tag{2.3.4}$$

i.e., $W(x, y)$ is zero if $x = y = 0$, otherwise, it is positive.

We can now use the function $W(x, y)$ to study the behavior of the solutions to Eq. (2.3.1). The system equations are

$$\dot{x} = y, \quad \dot{y} = -f(x) - \epsilon g(x)\dot{x} \tag{2.3.5}$$

and

$$\frac{dW}{dt} = \frac{\partial W}{\partial x}\dot{x} + \frac{\partial W}{\partial y}\dot{y}.$$

Substituting the expressions for $\dot{x}$ and $\dot{y}$, from Eq. (2.3.5), into the right-hand side gives

$$\frac{dW}{dt} = f(x)y + y[-f(x) - \epsilon g(x)y] = -\epsilon g(x)y^2 \leq 0. \tag{2.3.6}$$

The above inequality implies that W is a monotonic decreasing function and this in turn implies that

$$\operatorname*{Lim}_{t\to\infty} \begin{pmatrix} x(t) \\ y(t) \end{pmatrix} = \begin{pmatrix} 0 \\ 0 \end{pmatrix}. \tag{2.3.7}$$

Therefore, if the solution for $\epsilon = 0$ is periodic, then for $\epsilon > 0$, the periodic behavior is replaced by damped oscillatory motion.

Section 1.5 of the book by Jordan and Smith [7] presents several examples and further explanations of nonlinear damping.

2.3.1 *Damped Linear Oscillator*

The differential equation for this oscillator is

$$\ddot{x} + x = -2\epsilon\dot{x}, \quad \epsilon > 0, \tag{2.3.8}$$

and the function $W(x, y)$ is

$$W(x, y) = x^2 + y^2. \tag{2.3.9}$$

Using the system equations

$$\dot{x} = y, \quad \dot{y} = -x - 2\epsilon y,$$

we can conclude that all the solutions to the damped linear oscillatory go to zero as $t \to \infty$. This result can be easily checked since the exact solution for Eq. (2.3.8) is

$$x(t) = A_1 e^{-\epsilon t} \cos\left[\sqrt{1-\epsilon^2}t + \phi\right], \tag{2.3.10}$$

where A_1 and ϕ are arbitrary constants.

2.3.2 *Damped TNL Oscillator*

Consider the following TNL oscillator differential equation

$$\ddot{x} + x^{1/3} = -\epsilon x^2 (\dot{x})^3. \tag{2.3.11}$$

The damping term for this case is more general than the form given in Eq. (2.3.1), i.e.,

$$g(x) \to g(x, \dot{x}) = x^2 (\dot{x})^3. \tag{2.3.12}$$

The $W(x, y)$ function is

$$W(x, y) = \frac{y^2}{2} + \left(\frac{3}{4}\right) x^{4/3},$$

and the system equations are

$$\dot{x} = y, \quad \dot{y} = -x^{1/3} - \epsilon x^2 y^3. \tag{2.3.13}$$

Therefore

$$\frac{dW}{dt} = (x^{1/3})y + y\left[-x^{1/3} - \epsilon x^2 y^3\right] = -\epsilon x^2 y^4 \le 0. \tag{2.3.14}$$

Since for $\epsilon = 0$, periodic solutions exist, then for $\epsilon > 0$, the solutions become damped and oscillatory.

Figure 2.3.1 illustrates the generic cases for $\epsilon = 0$ and $\epsilon > 0$.

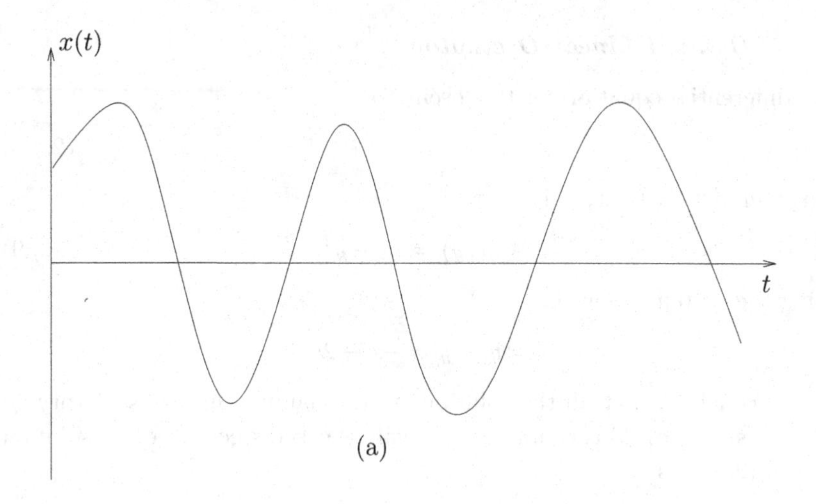

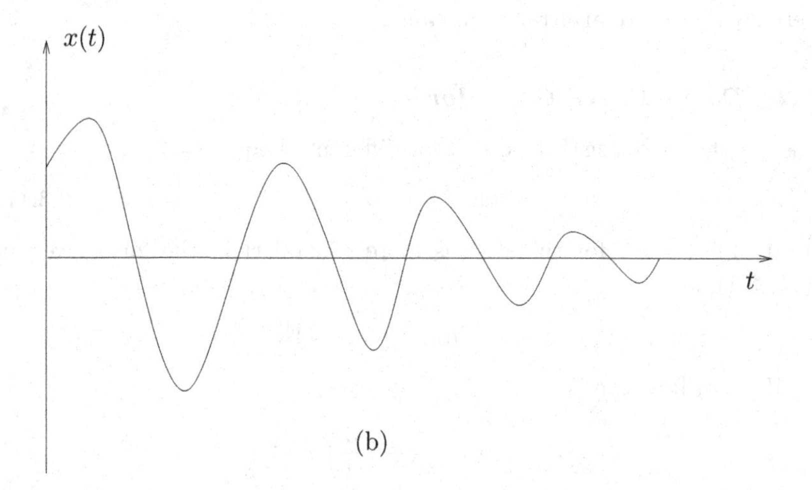

Fig. 2.3.1 $\ddot{x}+f(x)=-\epsilon g(x)\dot{x}$: (a) Periodic solutions for $\epsilon=0$. (b) Damped oscillatory solutions for $\epsilon>0$.

2.3.3 *Mixed-Damped TNL Oscillator*

An example of a TNL oscillator having an interesting damping term is

$$\ddot{x}+(1+\dot{x})x^{1/3}=0. \tag{2.3.15}$$

The damping is expressed by the term $\dot{x}x^{1/3}$. For this case the sign of the damping depends on the sign of $x^{1/3}$, i.e., comparison of Eq. (2.3.15) with Eqs. (2.3.1), (2.3.2), the $g(x)$ for this particular equation is $x^{1/3}$.

Inspection of Eq. (2.3.15) indicates that it is not invariant under $t \to -t$ or $x \to -x$, i.e., it is not invariant under time reversal or of odd-parity.

The system equations are

$$\dot{x} = y, \quad \dot{y} = -(1+y)x^{1/3}, \tag{2.3.16}$$

and the corresponding fixed-point is $(\bar{x}, \bar{y}) = (0, 0)$.

The differential equation for the trajectories in phase-space is

$$\frac{dy}{dx} = -\frac{(1+y)x^{1/3}}{y}, \tag{2.3.17}$$

and it is invariant only under the symmetry transformation

$$S_1 : x \to -x, \quad y \to y. \tag{2.3.18}$$

Since Eq. (2.3.17) is separable, it may be integrated to obtain the first-integral,

$$y - \ln|1+y| + \left(\frac{3}{4}\right) x^{4/3} = \left(\frac{3}{4}\right) A^{4/3}, \tag{2.3.19}$$

where the initial conditions $x(0) = A$, $\dot{x}(0) = y(0) = 0$, were used to determine the constant of integration.

From Eq. (2.3.17), it follows that the null-clines are

$y' = 0 : y_0(x)$ is the y-axis and the line $y = -1$.

$y' = \infty : y_\infty(x)$ is the x-axis.

For this TNL oscillator, $W(x, y)$ can be taken as

$$W(x,y) = \frac{y^2}{2} + \left(\frac{3}{4}\right) x^{4/3}, \tag{2.3.20}$$

and from this follows the result

$$\frac{dW}{dt} = -y^2 x^{1/3} : \begin{cases} > 0, & \text{for } x < 0; \\ 0, & \text{for } x = 0 \text{ or } y = 0; \\ < 0, & \text{for } x > 0. \end{cases}$$

Using all of the above information, the basic features of the phase-plane can be determined and these results are displayed in Figure 2.3.2. Two typical trajectories are shown in Figure 2.3.3. In summary, the solutions to the mixed-damped TNL oscillator, expressed by Eq. (2.3.15), have the following properties:

(i) For initial conditions

$$-\infty < x(0) < \infty, \quad y(0) > -1,$$

all the solutions, $x(t)$ and $y(t)$, are periodic.

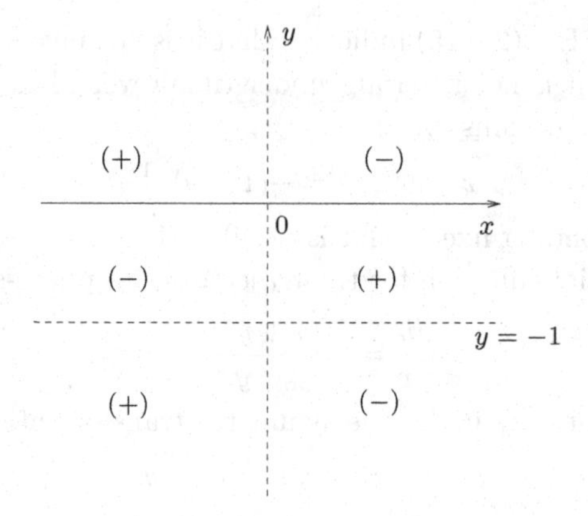

Fig. 2.3.2 Phase-plane for Eq. (2.3.15). The dashed lines are the $y_0(x)$ null-clines. The solid line is the $y_\infty(x)$ null-cline.

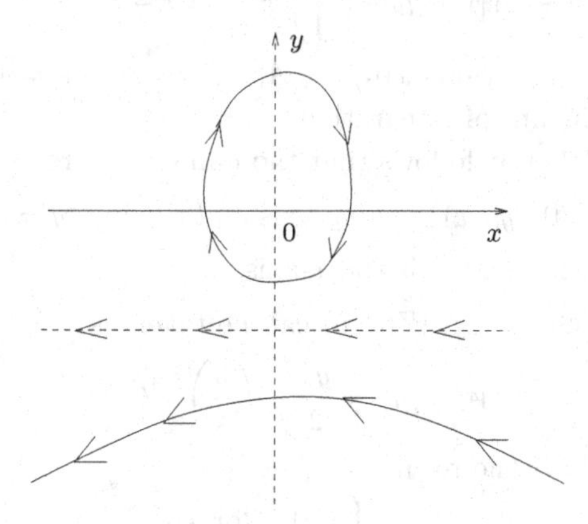

Fig. 2.3.3 Typical trajectories in the phase-plane for Eq. (2.3.17).

(ii) $y(t) = -1$ or $x(t) = x(0) - t$, is an exact solution to the differential equation.

(iii) If $y(0) < -1$, $x(0)$ arbitrary, then no periodic solutions can occur. These solutions become unbounded and eventually lie in the third quadrant of the phase-plane.

The general conclusion is that for a mixed-damped oscillator equation, some solutions may be periodic, other not. The type of solution obtained is dependent on the initial conditions.

2.4 Resume

A TNL oscillator differential equation modeling a conservative system has the form

$$\ddot{x} + f(x) = 0, \tag{2.4.1}$$

where $f(x)$ contains one or more terms having the structure

$$f(x) = \begin{cases} x^{\frac{2n+1}{2m+1}}, & (n, m) \text{ are integers;} \\ |x|^p \text{sgn}(x), & p = \text{real.} \end{cases} \tag{2.4.2}$$

The initial conditions are generally selected as

$$x(0) = A, \quad \dot{x}(0) = 0. \tag{2.4.3}$$

In spite of the name, i.e., TNL oscillator differential equations, prior arguments must be given to demonstrate that periodic solutions exist. The work of this chapter indicates that two methods exist for carrying out this task: the use of qualitative methods based on examining the geometrical properties of the trajectories in the 2-dim phase-space, and the use of first-integrals [4, 5, 15]. The first technique is more powerful since it may be applied in all situations. In either case, the goal is to show that either all or some of the trajectories in the phase-plane are closed. Since closed trajectories correspond to periodic solutions, the existence of periodic solutions is then established.

Related to the function $f(x)$ is the potential function, $V(x)$, defined as

$$V(x) \equiv \int f(x)dx. \tag{2.4.4}$$

The functions, $f(x)$, appearing in this volume will generally have properties such that if $V(0)$ exists, then $V(0) = 0$, and $V(x)$ is monotonic increasing.

Finally, after the task of demonstrating that periodic solutions exist, the next step is to create or construct analytical techniques for calculating approximations to these solutions. The remainder of the volume is concerned with this issue.

Problems

2.1 What are the system equations for the following TNL oscillator differential equations?

(i) $\ddot{x} + x + x^{1/3} = 0$

(ii) $\ddot{x} + \frac{x^3}{1+x^2} = 0$
(iii) $\ddot{x} + x^{1/3} = \epsilon(1 - x^2)\dot{x}$
(iv) $\ddot{x} + \frac{1}{x} = 0$
(v) $\ddot{x} + (1 + x^2 + \dot{x}^2)x^{1/3} = 0$.

2.2 Carry out a complete phase-plane analysis for each of the differential equations given in problem 2.1. Which of these equations have potential functions? If such functions exist, then calculate them and plot $V(x)$ vs x.

2.3 Given a TNL oscillator equation, show (by an explicit example) that invariance under $t \to -t$ and $x \to -x$ does not imply that periodic solutions exist.

2.4 Derive Eq. (2.2.12).

2.5 Reproduce the steps from Eq. (2.2.13) to Eq. (2.2.15).

2.6 For $W(x, y)$, as given in Eq. (2.3.3), show that $dW/dt \leq 0$, implies the result of Eq. (2.3.7).

2.7 Derive the properties of the phase-plane as shown in Figure 2.3.2 for the mixed-damped TNL oscillator differential equation

$$\ddot{x} + (1 + \dot{x})x^{1/3} = 0.$$

2.8 For problem 2.7, show that $x(t) = x_0 - t$ is an exact solution.

2.9 For periodic solutions of the mixed-damped TNL oscillator, calculate the locations of the two points of intersection of a closed trajectory with the y-axis. Hint: Select the initial values $x(0) = A > 0$ and $\dot{x}(0) = y(0) = 0$, and apply the first-integral.

2.10 What is the potential function associated with

$$\ddot{x} + \frac{1}{x^{1/3}} = 0?$$

Show that all trajectories in the phase-plane are bounded.

References

[1] E. A. Coddington and N. Levison, *Theory of Differential Equations* (McGraw-Hill, New York, 1995).
[2] J. K. Hale, *Oscillations in Nonlinear Systems* (McGraw-Hill, New York, 1963).
[3] S. L. Ross, *Nonlinear Differential Equations* (Blaisdell; Waltham, MA; 1964).
[4] G. Sansone and R. Conti, *Nonlinear Differential Equations* (Pergamon, New York, 1964).

[5] R. E. Mickens, *Oscillations in Planar Dynamic Systems* (World Scientific, London, 1996).
[6] L. Edelstein-Keshet, *Mathematical Models in Biology* (McGraw-Hill, New York, 1988). See Chapter 5. Also, see Appendix I of ref. [5]
[7] D. W. Jordan and P. Smith, *Nonlinear Ordinary Differential Equations*, 2nd edition (Clarendon Press, Oxford, 1987).
[8] G. H. Duffing, *Theoretical Physics* (Houghton Mifflin, Boston, 1973).
[9] L. Meirovitch, *Elements of Vibration Analysis* (McGraw-Hill, New York, 1975).
[10] S. H. Strogatz, *Nonlinear Dynamics and Chaos with Applications to Physics, Chemistry and Engineering* (Addison Wesley; Reading, MA; 1994).
[11] P. N. V. Tu, *Dynamical Systems: An Introduction with Applications in Economics and Biology* (Springer-Verlag, Berlin, 1994, 2nd edition).
[12] I. S. Gradshteyn and I. M. Ryzhik, *Tables of Integrals, Series, and Products* (Academic Press, New York, 1980). See formula 3.139, #5.
[13] H. Hancock, *Elliptic Integrals* (Wiley, New York, 1917). See pp. 52.
[14] P. F. Byrd and M. S. Friedmann, *Handbook of Elliptic Integrals for Engineers and Physicists* (Springer-Verlag, Berlin, 1954).
[15] V. Nemytskii and V. Stepanov, *Qualitative Theory of Differential Equations* (Princeton University Press; Princeton, NJ; 1959).

Chapter 3

Harmonic Balance

The method of harmonic balance provides a general technique for calculating approximations to the periodic solutions of differential equations. It corresponds to a truncated Fourier series and allows for the systematic determination of the coefficients to the various harmonics and the angular frequency. The significance of the method is that it may be applied to differential equations for which the nonlinear terms are not small.

There are a number of formulations of the method of harmonic balance. Mickens' book [1, Section 4.1], includes a list of some of the relevant publications on this topic. A new approach, using a rational harmonic balance formulation, was introduced by Beléndez et al. [2]; they demonstrate the utility of the procedure by applying it to several nonlinear oscillatory systems. The mathematical foundations of harmonic balancing have been investigated by several individuals. The works of Borges et al. [3], Miletta [4], and Bobylev et al. [5] provide overviews to various issues concerning convergence and error bounds for the approximations to the periodic solutions.

In Section 3.1, we present the methodology for the direct harmonic balance procedure and demonstrate its use by applying it to several examples of TNL oscillators in Section 3.2. Section 3.3 introduces a rational formulation of harmonic balancing and this is followed by Section 3.4, in which four TNL oscillators have approximations to their periodic solutions calculated. Two third-order equations are studied in Section 3.5; they arise in the investigation of stellar oscillations [6]. Finally, in Section 3.6, we make general comments on and give a summary of the main features and conclusions reached in this chapter.

3.1 Direct Harmonic Balance: Methodology

The presentation in this section relies heavily on my previous work. In particular, see Mickens [1, Section 4.2.2]; Mickens [7, 8]; and Mickens [9, Section 8.8.1].

All of the TNL oscillator equations studied (except for those in Section 3.5 can be written

$$F(x, \dot{x}, \ddot{x}) = 0, \tag{3.1.1}$$

where $F(\cdots)$ is of odd-parity, i.e.,

$$F(-x, -\dot{x}, -\ddot{x}) = -F(x, \dot{x}, \ddot{x}). \tag{3.1.2}$$

A major consequence of this property is that the corresponding Fourier expansions of the periodic solutions only contain odd harmonics [10], i.e.,

$$x(t) = \sum_{k=1}^{\infty} \{A_k \cos[(2k-1)\Omega t] + B_k \sin[(2k-1)\Omega t]\}. \tag{3.1.3}$$

The N-th order harmonic balance approximation to $x(t)$ is the expression

$$x_N(t) = \sum_{k=1}^{N} \{\bar{A}_k^N \cos[(2k-1)\bar{\Omega}_N t] + \bar{B}_k^N \sin[(2k-1)\bar{\Omega}_N t]\}, \tag{3.1.4}$$

where $(\bar{A}_k^N, \bar{B}_k^N, \bar{\Omega}_N)$ are approximations to (A_k, B_k, Ω) for $k = 1, 2, \ldots, N$.

For the case of a conservative oscillator, Eq. (3.3.1) generally takes the form

$$\ddot{x} + f(x, \lambda) = 0, \tag{3.1.5}$$

where λ denotes the various parameters appearing in $f(x, \lambda)$ and $f(-x, \lambda) = -f(x, \lambda)$. The following initial conditions are selected

$$x(0) = A, \quad \dot{x}(0) = 0, \tag{3.1.6}$$

and this has the consequence that only the cosine terms are needed in the Fourier expansions, and therefore we have

$$x_N(t) = \sum_{k=1}^{N} \bar{A}_k^N \cos[(2k-1)\Omega_N t]. \tag{3.1.7}$$

Observe that $x_N(t)$ has $(N+1)$ unknowns, the N coefficients, $(\bar{A}_1^N, \bar{A}_2^N, \ldots, \bar{A}_N^N)$ and Ω_N, the angular frequency. These quantities may be calculated by carrying out the following steps:

(1) Substitute Eq. (3.1.7) into Eq. (3.1.5), and expand the resulting form into an expression that has the structure

$$\sum_{k=1}^{N} H_k \cos[(2k-1)\Omega_N t] + \text{HOH} \simeq 0, \tag{3.1.8}$$

where the H_m are functions of the coefficients, the angular frequency, and the parameters, i.e.,

$$H_k = H_k(\bar{A}_1^N, \bar{A}_2^N, \ldots, \bar{A}_N^N, \Omega_N, \lambda).$$

Note that in Eq. (3.1.8), we only retain as many harmonics in our expansion as initially occur in the assumed approximation to the periodic solution.

(2) Set the functions H_k to zero, i.e.,

$$H_k = 0, \quad k = 1, 2, \ldots, N. \tag{3.1.9}$$

The action is justified because the cosine functions are linearly independent and, as a result, any linear sum of them that is equal to zero must have the property that the coefficients are all zero; see Mickens [9, pp. 221–222].

(3) Solve the N equations, see Eq. (3.1.9), for $\bar{A}_2^N, \bar{A}_3^N, \ldots, \bar{A}_N^N$ and Ω_N, in terms of $\bar{A}_1^N$.

Using the initial conditions, Eq. (3.1.6), we have for $\bar{A}_1^N$ the relation

$$x_N(0) = A = \bar{A}_1^N + \sum_{k=2}^{N} \bar{A}_k^N(\bar{A}_1^N, \lambda). \tag{3.1.10}$$

An important point is that Eq. (3.1.9) will have many distinct solutions and the "one" selected for a particular oscillator equation is that one for which we have known *a priori* restrictions on the behavior of the approximations to the coefficients. However, as the worked examples in the next section demonstrate, in general, no essential difficulties arise.

For nonconservative oscillators, where $\dot{x}$ appears to an "odd power," the calculation of approximations to periodic solutions follows a procedure modified for the case of conservative oscillators presented above. Many of these equations take the form

$$\ddot{x} + f(x, \lambda_1) = g(x, \dot{x}, \lambda_2)\dot{x}, \tag{3.1.11}$$

where

$$f(-x, \lambda_1) = -f(x, \lambda_1), \quad g(-x, -\dot{x}, \lambda_2) = g(x, \dot{x}, \lambda_2), \tag{3.1.12}$$

and (λ_1, λ_2) denote the parameters appearing in f and g. For this type of differential equation, a limit-cycle may exist and the initial conditions cannot, in general, be *a priori* specified [1, 11, 12]. (Limit-cycles correspond to isolated closed curves in the phase-place; see [1, Appendix G].)

Harmonic balancing, for systems where limit-cycles may exist, uses the following procedures:

(1) Take the N-th order approximation to the periodic solution to be

$$x_N(t) = \bar{A}_1^N \cos(\bar{\Omega}_N t) + \sum_{k=2}^{N} \{\bar{A}_k^N \cos[(2k-1)\bar{\Omega}_N t] + \bar{B}_k^N \sin[(2k-1)\bar{\Omega}_N t]\}, \quad (3.1.13)$$

where the $2N$ unknowns

$$\begin{cases} \bar{A}_1^N, \bar{A}_2^N, \ldots, \bar{A}_N^N; \\ \bar{\Omega}_N, \bar{B}_2^N, \ldots, \bar{B}_N^N; \end{cases} \quad (3.1.14)$$

are to be determined.

(2) Substitute Eq. (3.1.13) into Eq. (3.1.11) and write the result as

$$\sum_{k=1}^{N} \{H_k \cos[(2k-1)\Omega_N T] + L_k \sin[(2k-1)\Omega_N t]\} + \text{HOH} \simeq 0, \quad (3.1.15)$$

where the $\{H_k\}$ and $\{L_k\}$, $k = 1$ to N, are functions of the $2N$ unknowns listed in Eq. (3.1.14).

(3) Next equate the $2N$ functions $\{H_k\}$ and $\{L_k\}$ to zero and solve them for the $(2N-1)$ amplitudes and the angular frequency. If a "valid" solution exists, then it corresponds to a limit-cycle. In general, the amplitudes and angular frequency will be expressed in terms of the parameters λ_1 and λ_2.

As stated earlier in this section, the method of harmonic balance may give spurious solutions. Therefore, one should obtain prior knowledge from the use of other procedures, such as a phase-plane analysis, to insure that correct solutions are derived from the application of this method. Another criterion is to require that the approximate Fourier coefficients satisfy relevant bounds on their values as a function of their index label, k; see [1, Section C.3].

3.2 Worked Examples

We illustrate the application and effectiveness of the direct harmonic balance method by using it to determine approximations to the periodic solution to five TNL oscillators. In each case, we calculate the second approximation to demonstrate the technique. Higher order expressions merely require more algebraic manipulations and effort.

3.2.1 $\ddot{x} + x^3 = 0$

We begin by calculating the first-order harmonic balance approximation to the periodic solutions of

$$\ddot{x} + x^3 = 0, \quad x(0) = A, \quad \dot{x}(0) = 0. \tag{3.2.1}$$

This approximation takes the form

$$x_1(t) = A\cos(\Omega_1 t). \tag{3.2.2}$$

Observe that this expression automatically satisfies the initial conditions. Substituting Eq. (3.2.2) into Eq. (3.2.1) gives ($\theta = \Omega_1 t$)

$$(-A\Omega_1^2 \cos\theta) + (A\cos\theta)^3 \simeq 0,$$

$$-(A\Omega_1^2)\cos\theta + A^3\left[\left(\frac{3}{4}\right)\cos\theta + \left(\frac{1}{4}\right)\cos 3\theta\right] \simeq 0,$$

$$A\left[-\Omega_1^2 + \left(\frac{3}{4}\right)A^2\right]\cos\theta + \text{HOH} \simeq 0.$$

Setting the coefficient of $\cos\theta$ to zero gives the first-approximation to the angular frequency

$$\Omega_1(A) = \left(\frac{3}{4}\right)^{1/2} A, \tag{3.2.3}$$

and

$$x_1(t) = A\cos\left[\left(\frac{3}{4}\right)^{1/2} At\right]. \tag{3.2.4}$$

The solution for the second-approximation takes the form ($\theta = \Omega_2 t$)

$$x_2(t) = A_1\cos\theta + A_2\cos 3\theta \tag{3.2.5}$$

with

$$\ddot{x}_2(t) = -\Omega_2^2(A_1\cos\theta + 9A_2\cos 3\theta). \tag{3.2.6}$$

Substituting Eq. (3.2.5) and Eq. (3.2.6) into Eq. (3.2.1) gives

$$H_1(A_1, A_2, \Omega_2)\cos\theta + H_2(A_1, A_2, \Omega_2)\cos 3\theta + \text{HOH} \simeq 0,$$

where

$$H_1 = A_1\left[\Omega_2^2 - \left(\frac{3}{4}\right)A_1^2 - \left(\frac{3}{4}\right)A_1 A_2 - \left(\frac{3}{2}\right)A_2^2\right], \tag{3.2.7}$$

$$H_2 = -9A_2\Omega_2^2 + \left(\frac{1}{4}\right)A_1^3 + \left(\frac{3}{2}\right)A_1^2A_2 + \left(\frac{3}{4}\right)A_2^3. \tag{3.2.8}$$

Setting H_1 to zero, and defining z as

$$z \equiv \frac{A_2}{A_1}, \tag{3.2.9}$$

we obtain

$$\Omega_2 = \left(\frac{3}{4}\right)^{1/2} A_1(1+z+2z^2)^{1/2} = \Omega_1(1+z+2z^2)^{1/2}, \tag{3.2.10}$$

where Ω_1 is that of Eq. (3.2.3). Inspection of Eq. (3.2.10) shows that the second approximation for the angular frequency is a modification of the first-approximation result.

If this value for Ω_2 is substituted into Eq. (3.2.8) and this expression is set to zero, and if the definition of z is used, then the following cubic equation must be satisfied by z

$$51z^3 + 27z^2 + 21z - 1 = 0. \tag{3.2.11}$$

There are three roots, but the one of interest should be real and have a small magnitude, i.e.,

$$|z| \ll 1. \tag{3.2.12}$$

This root is

$$z_1 = 0.044818\ldots, \tag{3.2.13}$$

and implies that the amplitude, A_2, of the higher harmonic, i.e., the $\cos 3\theta$, is less than 5% of the amplitude of the fundamental mode, $\cos\theta$.

Therefore, the second harmonic balance approximation for Eq. (3.2.1) is

$$x_2(t) = A_1[\cos\theta + z_1\cos 3\theta].$$

For the initial condition, $x_2(0) = A$, we find

$$A = A_1(1+z_1) \quad \text{or} \quad A_1 = \frac{A}{1+z_1} = (0.9571)A. \tag{3.2.14}$$

Using this result in Eq. (3.2.10) gives

$$\Omega_2(A) = \left(\frac{3}{4}\right)^{1/2} A\left[\frac{(1+z_1+2z_1^2)^{1/2}}{1+z_1}\right] = (0.8489)A. \tag{3.2.15}$$

The corresponding periods ($T = 2\pi/\Omega$) are

$$T_1 \equiv \frac{2\pi}{\Omega_1} = \frac{7.2554}{A}, \quad T_2 \equiv \frac{2\pi}{\Omega_2} = \frac{7.4016}{A}, \tag{3.2.16}$$

$$T_{\text{exact}} = \frac{7.4163}{A}, \tag{3.2.17}$$

and they have the following percentage errors

$$E_1(\%) = 2.2\% \qquad E_2(\%) = 0.20\% \tag{3.2.18}$$

where

$$E \equiv \text{percentage error} = \left| \frac{T_{\text{exact}} - T}{T_{\text{exact}}} \right| \cdot 100. \tag{3.2.19}$$

Since the differential equation Eq. (3.2.1) has the exact solution [1]

$$x(t) = A\, cn\left(At; 1/\sqrt{2}\right),$$

where "*cn*" is the Jacobi cosine function the ratio, A_2/A_1, can be explicitly calculated; its value is 0.045078. This should be compared to our value of 0.044818.

In summary, the second-order harmonic balance approximation for the periodic solution of Eq. (3.2.1) is

$$x_2(t) = \left(\frac{A}{1+z_1}\right)[\cos(\Omega_2 t) + z_1 \cos(3\Omega_2 t)], \tag{3.2.20}$$

where z_1 and Ω_2 are given, respectively, in Eqs. (3.2.13) and (3.2.15).

3.2.2 $\ddot{x} + x^{-1} = 0$

The above differential equation was studied by Mickens [14] and occurs as a model of certain phenomena in plasma physics [15]. Note that Acton and Squire [15] give an elegant, but simple algebraic approximation to the periodic solution of this equation

$$\ddot{x} + \frac{1}{x} = 0, \quad x(0) = A, \quad \dot{x}(0) = 0. \tag{3.2.21}$$

The exact period can be calculated and its value is

$$T_{\text{exact}}(A) = 2\sqrt{2}A \int_0^1 \frac{ds}{\sqrt{\ln\left(\frac{1}{s}\right)}} = 2\sqrt{2\pi}\, A, \tag{3.2.22}$$

where the value of the integral is given in Gradshteyn and Ryzhik [16]. The corresponding angular frequency is

$$\Omega_{\text{exact}}(A) = \frac{2\sqrt{2\pi}}{2A} = \frac{1.2533141}{A}. \tag{3.2.23}$$

For the first-order harmonic balance, the solution is $x_1(t) = A\cos\theta$, $\theta = \Omega_1 t$. This calculation is best achieved if Eq. (3.2.21) is rewritten to the form

$$x\ddot{x} + 1 = 0. \tag{3.2.24}$$

Substituting $x_1(t)$ into this equation gives

$$(A\cos\theta)(-\Omega_1^2 A\cos\theta) + 1 + \text{HOH} \simeq 0,$$

or

$$\left[-\left(\frac{\Omega_1^2 A^2}{2}\right) + 1\right] + \text{HOH} \simeq 0. \tag{3.2.25}$$

Therefore, in lowest order, the angular frequency is

$$\Omega_1(A) = \frac{\sqrt{2}}{A} = \frac{1.4142}{A}. \tag{3.2.26}$$

The second harmonic balance approximation is

$$x_2(t) = A_1\cos\theta + A_2\cos 3\theta, \quad \theta = \Omega_2 t. \tag{3.2.27}$$

Substituting this expression into Eq. (3.2.24) gives

$$(A_1\cos\theta + A_2\cos 3\theta)[-\Omega_2^2(A_1\cos\theta + 9A_2\cos 3\theta)] + 1 \simeq 0,$$

and on performing the required expansions, we obtain

$$\left[-\Omega_2^2\left(\frac{A_1^2 + 9A_2^2}{2}\right) + 1\right] - \Omega_2^2\left(\frac{A_1^2 + 10A_1A_2}{2}\right)\cos 2\theta + \text{HOH} \simeq 0.$$

Setting the constant term and the coefficient of $\cos 2\theta$ to zero gives

$$-\Omega_2^2\left(\frac{A_1^2 + 9A_2^2}{2}\right) + 1 = 0, \quad A_1^2 + 10A_1A_2 = 0, \tag{3.2.28}$$

with the solutions

$$A_2 = -\left(\frac{A_1}{10}\right), \quad \Omega_2^2 = \frac{200}{109A_1^2}. \tag{3.2.29}$$

Therefore,

$$x_2(t) = A_1\left[\cos(\Omega_2 t) - \left(\frac{1}{10}\right)\cos(3\Omega_2 t)\right],$$

and requiring

$$x_2(0) = A = \left(\frac{9}{10}\right)A_1 \quad \text{or} \quad A_1 = \left(\frac{10}{9}\right)A,$$

gives

$$x_2(t) = \left(\frac{10}{9}\right) A \left[\cos(\Omega_2 t) - \left(\frac{1}{10}\right)\cos(3\Omega_2 t)\right], \tag{3.2.30}$$

with

$$\Omega_2^2 = \frac{200}{109A_1^2} = \left(\frac{162}{109}\right)\frac{1}{A^2}$$

or

$$\Omega_2(A) = \frac{1.2191138}{A}. \tag{3.2.31}$$

The percentage error is

$$\left|\frac{\Omega_{\text{exact}} - \Omega_2}{\Omega_{\text{exact}}}\right| \cdot 100 = 2.7\% \text{ error.}$$

Note that the first approximation gives

$$\left|\frac{\Omega_{\text{exact}} - \Omega_1}{\Omega_{\text{exact}}}\right| \cdot 100 = 12.8\% \text{ error.}$$

3.2.3 $\ddot{x} + x^2 sgn(x) = 0$

The quadratic oscillator is modeled by the equation

$$\ddot{x} + x^2 \text{sgn}(x) = 0. \tag{3.2.32}$$

To apply the harmonic balance method, we rewrite it to the form

$$(\ddot{x})^2 = x^4, \tag{3.2.33}$$

using

$$[\text{sgn}(x)]^2 = 1.$$

For first order harmonic balance, where $x_1(t) = A\cos\theta$, with $\theta = \Omega_1 t$, we have

$$[-A\Omega_1^2 \cos\theta]^2 \simeq [A\cos\theta]^4. \tag{3.2.34}$$

Using

$$(\cos\theta)^2 = \left(\frac{1}{2}\right) + \left(\frac{1}{2}\right)\cos 2\theta,$$

$$(\cos\theta)^4 = \left(\frac{3}{8}\right) + \left(\frac{1}{2}\right)\cos 2\theta + \left(\frac{1}{8}\right)\cos 4\theta,$$

we obtain from Eq. (3.2.34)

$$\left(\frac{A^2\Omega_1^4}{2}\right) \simeq \left(\frac{3A^4}{8}\right) + \text{HOH}, \tag{3.2.35}$$

which gives the following result for Ω_1

$$\Omega_1^{(1)}(A) = \left(\frac{3}{4}\right)^{1/4} A^{1/2}. \tag{3.2.36}$$

The reason we write $\Omega_1^{(1)}(A)$, rather than $\Omega_1(A)$, is that a second version of the simple harmonic balance approximation can be derived; see Section 4.3.6 of [1] and [18].

Equation (3.2.32) can also be written

$$\ddot{x} + |x|x = 0. \tag{3.2.37}$$

Using the result [18]

$$|\cos\theta| = \left(\frac{4}{\pi}\right)\left[\left(\frac{1}{2}\right) + \left(\frac{1}{3}\right)\cos 2\theta - \left(\frac{1}{15}\right)\cos 4\theta + \cdots\right],$$

then the first harmonic balance approximation becomes

$$-\left[\Omega_1^{(2)}\right]^2 A\cos\theta + A^2|\cos\theta|\cos\theta \simeq 0,$$

or

$$\left\{-\left[\Omega_1^{(2)}\right]^2 + \frac{8A}{3\pi}\right\} A\cos\theta + \text{HOH} \simeq 0. \tag{3.2.38}$$

Setting the coefficient of $\cos\theta$ to zero gives

$$\Omega_1^{(2)}(A) = \left(\frac{8}{3\pi}\right)^{1/2} A^{1/2}. \tag{3.2.39}$$

For comparison, we have

$$\Omega_1^{(1)}(A) = (0.93060\ldots)A^{1/2}, \quad \Omega_1^{(2)}(A) = (0.92131\ldots)A^{1/2}. \tag{3.2.40}$$

To calculate the second-order harmonic balance solution, we use Eq. (3.2.33) and take

$$x_2(t) = A_1\cos\theta + A_2\cos 3\theta.$$

If we define

$$z \equiv \frac{A_2}{A_1}, \tag{3.2.41}$$

then

$$\begin{cases} x_2(t) = A_1[\cos\theta + z\cos 3\theta], \\ \ddot{x}_2(t) = -\Omega_2^2 A_1[\cos\theta + 9z\cos 3\theta]. \end{cases} \tag{3.2.42}$$

A straightforward, but long calculation gives

$$(\ddot{x}_2)^2 = \left(\frac{\Omega_2^4 A_1^2}{2}\right)\left[(1+81z^2) + (1+18z)\cos 2\theta + \text{HOH}\right], \tag{3.2.43}$$

$$\begin{aligned} x_2^4 = {} & A_1^4\left[\left(\frac{3}{8}\right) + \left(\frac{1}{2}\right)z + \left(\frac{3}{2}\right)z^2 + \left(\frac{3}{8}\right)z^4\right] \\ & + A_1^4\left[\left(\frac{1}{2}\right) + \left(\frac{3}{2}\right)z + \left(\frac{3}{2}\right)z^2 + \left(\frac{3}{2}\right)z^3\right]\cos 2\theta \\ & + \text{HOH}. \end{aligned} \tag{3.2.44}$$

Harmonic balancing, i.e., equating the coefficients of the constant and $\cos 2\theta$ terms in the last two equations, gives

$$\left(\frac{\Omega_2^4 A_1^2}{2}\right)(1+81z^2) = \left[\left(\frac{3}{8}\right) + \left(\frac{1}{2}\right)z + \left(\frac{3}{2}\right)z^2 + \left(\frac{3}{8}\right)z^4\right]A_1^4 \tag{3.2.45}$$

$$\left(\frac{\Omega_2^4 A_1^2}{2}\right)(1+18z) = \left[\left(\frac{1}{2}\right) + \left(\frac{3}{2}\right)z + \left(\frac{3}{2}\right)z^2 + \left(\frac{3}{2}\right)z^3\right]A_1^4. \tag{3.2.46}$$

Dividing these two expressions and simplifying the resulting expression yields a single equation for z,

$$(243)z^5 + \left(\frac{915}{4}\right)z^4 + (192)z^3 + (63)z^2 - \left(\frac{23}{2}\right)z + \frac{1}{4} = 0. \tag{3.2.47}$$

The smallest (in magnitude) real root is

$$z = 0.025627. \tag{3.2.48}$$

If Eq. (3.2.45) is solved for Ω_2, the following result is obtained

$$\Omega_2(A_1) = \left[\left(\frac{3}{4}\right)A_1^2\right]^{1/4} \cdot \left[\frac{1+\left(\frac{4}{3}\right)z + 4z^2 + z^4}{1+81z}\right]^{1/4}. \tag{3.2.49}$$

Since $x(0) = A = A_1(1+z)$, we have

$$\Omega_2(A) = \left\{\left(\frac{3}{4}\right)\left[\frac{1+\left(\frac{4}{3}\right)z + 4z^2 + z^4}{1+2z+z^2}\right]\right\}^{1/4} A^{1/2} = (0.927244)A^{1/2}. \tag{3.2.50}$$

In summary, the harmonic balance approximation for the periodic solution to the quadratic oscillator is

$$x_2(t) = \left(\frac{A}{1+z}\right)[\cos(\Omega_2 t) + z\cos(3\Omega_2 t)], \tag{3.2.51}$$

where z and $\Omega_2(A)$ have the values, respectively, given in Eqs. (3.2.48) and (3.2.50).

3.2.4 $\ddot{x} + x^{1/3} = 0$

The "cube-root" oscillator

$$\ddot{x} + x^{1/2} = 0, \quad x(0) = A, \quad \dot{x}(0) = 0, \tag{3.2.52}$$

was one of the first TNL oscillator equations to be investigated [19–21]. Its period can be calculated exactly [21].

The system equations for this oscillator are

$$\frac{dx}{dt} = y, \quad \frac{dy}{dt} = -x^{1/3}, \tag{3.2.53}$$

and the differential equation for the trajectories in the phase-plane, (x, y), is

$$\frac{dy}{dx} = -\frac{x^{1/3}}{y}, \tag{3.2.54}$$

and therefore a first-integral is

$$\frac{y^2}{2} + \left(\frac{3}{4}\right) x^{4/3} = \left(\frac{3}{4}\right) A^{4/3}. \tag{3.2.55}$$

From this the period of the oscillation can be determined by the following relation [1, 21]

$$T(A) = \sqrt{\frac{32}{3}} \int_0^A \frac{dx}{\sqrt{A^{4/3} - x^{4/3}}}. \tag{3.2.56}$$

With the change of variable, $x = Aw^{3/2}$, we find after some simplification

$$T(A) = \left(2\sqrt{6}\right) \phi A^{1/3}, \tag{3.2.57}$$

where

$$\phi \equiv \int_0^1 \sqrt{\frac{w}{(1+w)(1-w)}} dw. \tag{3.2.58}$$

Using Gradshteyn and Ryzhik [16], see Section 3.14 (formula 10), ϕ is

$$\phi = 2\sqrt{2}E\left(\frac{\pi}{2}, \frac{1}{\sqrt{2}}\right) - \sqrt{2}F\left(\frac{\pi}{2}, \frac{1}{\sqrt{2}}\right),$$

where "F" and "E" are, respectively, complete elliptic integrals of the first and second kinds [9, 16]. Using these results gives for the angular frequency the expression

$$\Omega_{\text{exact}}(A) = \frac{1.070451}{A^{1/3}}, \tag{3.2.59}$$

where $\Omega(A)T(A) = 2\pi$.

There are several ways in which a first-order harmonic balance approximation can be constructed. First, $x_1(t) = A\cos\theta$, $\theta = \Omega_1^{(1)}t$, can be used in the equation

$$\ddot{x} + x^{1/3} = 0,$$

to give

$$\left[-(\Omega_1^{(1)})^2 A\cos\theta\right] + (A\cos\theta)^{1/3} \simeq 0. \tag{3.2.60}$$

Using [9]

$$(\cos\theta)^{1/3} = a_1\cos\theta + \text{HOH}, \quad a_1 = 1.15959526, \tag{3.2.61}$$

Eq. (3.2.60) becomes

$$\left[-(\Omega_1^{(1)})^2 A + A^{1/3}a_1\right]\cos\theta + \text{HOH} \simeq 0,$$

and, for $\Omega_1^{(1)}$, we obtain

$$\Omega_1^{(1)} = \frac{\sqrt{a_1}}{A^{1/3}} = \frac{1.076844}{A^{1/3}}. \tag{3.2.62}$$

A second way to obtain a first-order harmonic balance approximation is to rewrite the differential equation to the following form

$$(\ddot{x})^3 + x = 0. \tag{3.2.63}$$

Substituting $x_1(t) = A\cos\theta$, $\theta = \Omega_1^{(2)}t$, into this equation gives

$$\left[-(\Omega_1^{(2)})^2 A\cos\theta\right]^3 + A\cos\theta \simeq 0,$$

and

$$A\left[1 - \left(\frac{3}{4}\right)A^2(\Omega_1^{(2)})^6\right]\cos\theta + \text{HOH} \simeq 0.$$

Therefore, for this case, the angular frequency is

$$\Omega_1^{(2)}(A) = \left(\frac{4}{3}\right)^{1/6}\left(\frac{1}{A^{1/3}}\right) = \frac{1.049115}{A^{1/3}}. \tag{3.2.64}$$

Comparing $\Omega_1^{(1)}(A)$ and $\Omega_1^{(2)}$ with Ω_{exact}, we obtain the following values for the percentage errors

$$E_1^{(1)} = 0.6\% \text{ error}, \quad E_1^{(2)} = 2.0\% \text{ error}. \tag{3.2.65}$$

Also observe that

$$\Omega_1^{(2)}(A) < \Omega_{\text{exact}}(A) < \Omega_1^{(1)}(A). \tag{3.2.66}$$

We now apply the second-order harmonic balance method to the cube-root equation expressed in the form of Eq. (3.2.63). For this case

$$x_2(t) = A_1[\cos\theta + z\cos 3\theta],$$

where $\theta = \Omega_2 t$, and $A_2 = zA_1$. Note that the second derivative is

$$\ddot{x}_2(t) = -(\Omega_2)^2 A[\cos\theta + 9z\cos 3\theta].$$

Substituting these two relations in Eq. (3.2.63) gives, after some algebraic and trigonometric simplification, the expression

$$\begin{aligned} &\left\{(\Omega_2)^6 A_1^3 \left[\left(\frac{3}{4}\right) + \left(\frac{27}{4}\right) z + \left(\frac{243}{2}\right) z^2\right] - A_1\right\}\cos\theta \\ &+ \left\{\Omega_2^6 A_1^3 \left[\left(\frac{1}{4}\right) + \left(\frac{2187}{4}\right) z^3 + \left(\frac{27}{2}\right) z\right] - zA_1\right\}\cos 3\theta \\ &+ \text{HOH} \simeq 0. \end{aligned} \tag{3.2.67}$$

Setting the coefficients of $\cos\theta$ and $\cos 3\theta$ to zero yields the following algebraic equations to be solved for z and Ω_2,

$$(\Omega_2)^6 A_1^2 \left[\left(\frac{3}{4}\right) + \left(\frac{27}{4}\right) z + \left(\frac{243}{2}\right) z^2\right] = 1,$$

$$(\Omega_2)^6 A_1^2 \left[\left(\frac{1}{4}\right) + \left(\frac{27}{4}\right) z + \left(\frac{2187}{4}\right) z^3\right] = z.$$

Dividing the two expressions gives a cubic equation to be solved for z, i.e.,

$$(1701)z^3 - (27)z^2 + (51)z + 1 = 0. \tag{3.2.68}$$

The smallest (in magnitude) real root of this equation is

$$z = -0.019178. \tag{3.2.69}$$

Since $A_1 = A/(1+z)$, the first equation allows the evaluation of the angular frequency, i.e.,

$$\Omega_2(A) = \left[\frac{1 + 2z + z^2}{\left(\frac{3}{4}\right) + \left(\frac{27}{4}\right) z + \left(\frac{243}{2}\right) z^2}\right]^{1/6} \cdot \left(\frac{1}{A^{1/3}}\right) = \frac{1.063410}{A^{1/3}}. \tag{3.2.70}$$

The percentage error in comparison with the exact value, $\Omega_{\text{exact}}(A)$, is

$$E_2 = 0.7\% \text{ error}. \tag{3.2.71}$$

Table 3.2.1 provides a summary of the results on $\Omega(A)$. As expected, second-order harmonic balance provides an improved value for the angular frequency in comparison with its associated first-order calculation.

Table 3.2.1 Values for $A^{1/3}\Omega(A)$.

Differential Equation	HB 1+	HB 2++
$\ddot{x} + x^{1/3} = 0$	1.076844	NA
$(\ddot{x})^3 + x = 0$	1.049115	1.063410

+First-order harmonic balance.
++Second-order harmonic balance.
$A^{1/3}\Omega_{\text{exact}}(A) = 1.070451$.

3.2.5 $\ddot{x} + x^{-1/3} = 0$ [22]

The above equation is the inverse-cube-root (ICR) oscillator and has several interesting features. Our goal is to present a complete discussion of this equation based on its known properties.

The two first-order system equations, corresponding to the ICR oscillator

$$\ddot{x} + \frac{1}{x^{1/3}} = 0, \quad x(0) = A, \quad \dot{x}(0) = 0, \tag{3.2.72}$$

are

$$\dot{x} = y, \quad \dot{y} = -\frac{1}{x^{1/3}}. \tag{3.2.73}$$

Note that these equations do not have any fixed points (constant solutions) in the finite (x, y) phase-plane.

The trajectories in the phase-plane, $y = y(x)$, are solutions to the following first-order differential equation

$$\frac{dy}{dx} = -\frac{1}{x^{1/3}y}, \tag{3.2.74}$$

and this separable equation can be solved to give a first-integral for Eq. (3.2.72)

$$\frac{y^2}{2} + \left(\frac{3}{2}\right)x^{2/3} = \left(\frac{3}{2}\right)A^{2/3}. \tag{3.2.75}$$

Inspection of Eq. (3.2.74) shows that the trajectory differential equation is invariant under the transformations S_1, S_2, and S_3 where

$$\begin{aligned} S_1 &: x \to -x, \quad y \to y, \\ S_2 &: x \to x, \quad\;\; y \to -y, \\ S_3 &: x \to -x, \quad y \to -y. \end{aligned} \tag{3.2.76}$$

Further examination of Eq. (3.2.74) indicates that there is only one null-cline, the one for which trajectories cross it with infinite slope. This null-cline, $y_\infty(x)$ consists of two segments and they coincide with the x and

y axes. As a consequence of this result, the phase-plane has four open domains, with the x and y axes being the boundaries. See Figure 3.2.1 for a representation of the phase-plane; the (+)/(−) denote the "sign" of dy/dx in the indicated open domain. Of critical importance is that the slope of the trajectory is infinite whenever the trajectory crosses a coordinate axis. The analytical expression of this trajectory curve is Eq. (3.2.75).

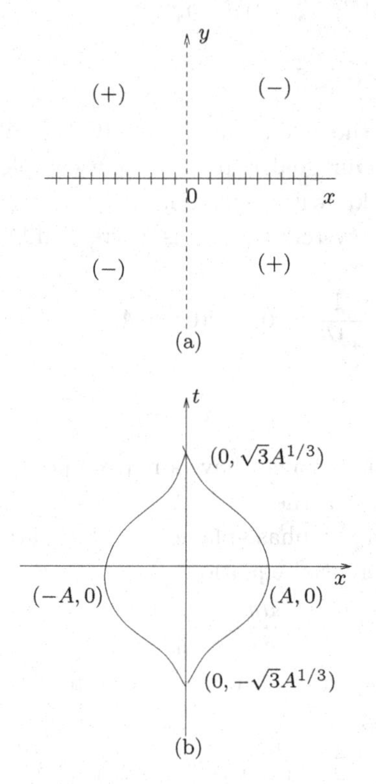

Fig. 3.2.1 (a) Phase-plane for $\ddot{x} + x^{-1/3} = 0$. Vertical dashes denote the infinite null-cline, $y_\infty(x)$. (b) Trajectory passing through $x(0) = A$ and $y(0) = 0$.

Since all of the trajectories are closed curves, then all solutions to the ICR differential equations are periodic.

An exact expression can be calculated for the period. To begin, the following relation holds for trajectories in the fourth quadrant,

$$y = \frac{dx}{dt} = -\sqrt{3(A^{2/3} - x^{2/3})}, \tag{3.2.77}$$

and this can be rewritten to the differential form

$$dt = -\left(\frac{1}{\sqrt{3}}\right)\frac{dx}{\sqrt{A^{2/3}-x^{2/3}}}.$$

Based on the symmetry transformations of Eq. (3.2.74) and an examination of Figure 3.2.1(b), it follows that

$$\int_0^{T/4} dt = -\left(\frac{1}{\sqrt{3}}\right)\int_A^0 \frac{dx}{\sqrt{A^{2/3}-x^{2/3}}}, \tag{3.2.78}$$

where $T = T(A)$ is the period of the oscillation. Replacing x by Az, Eq. (3.2.78) becomes

$$T(A) = \left(\frac{4}{\sqrt{3}}\right)A^{2/3}\int_0^1 \frac{dz}{\sqrt{1-z^{2/3}}}. \tag{3.2.79}$$

For $u = z^{2/3}$ change, the integral becomes

$$\int_0^1 \frac{dz}{\sqrt{1-z^{2/3}}} = \left(\frac{3}{2}\right)\int_0^1 \sqrt{\frac{u}{1-u}}\,du = \left(\frac{3}{2}\right)B\left(\frac{1}{2},\frac{3}{2}\right), \tag{3.2.80}$$

where $B(p,q)$ is the beta function [9]. Now $B\left(\frac{1}{2},\frac{3}{2}\right) = \frac{\pi}{2}$, and therefore

$$T(A) = \sqrt{3}\pi A^{2/3}$$

or

$$\Omega_{\text{exact}}(A) = \frac{2\pi}{T(A)} = \sqrt{\frac{4}{3}}\left(\frac{1}{A^{2/3}}\right) = \frac{1.1547005}{A^{2/3}}. \tag{3.2.81}$$

First-order harmonic balance solutions may be calculated by using one or the other of the equations

$$\begin{cases} x^{1/3}\ddot{x} + 1 = 0, \\ x(\ddot{x})^3 + 1 = 0. \end{cases} \tag{3.2.82}$$

For $x_1(t) = A\cos\theta$, $\theta = \Omega_1 t$, we obtain, respectively, the results

$$\Omega_1^{(1)}(A) = \left(\frac{2}{a_1}\right)^{1/2}\left(\frac{1}{A^{2/3}}\right) = \frac{1.3132934}{A^{2/3}}, \tag{3.2.83}$$

$$\Omega_1^{(2)}(A) = \left(\frac{8}{3}\right)^{1/6}\left(\frac{1}{A^{2/3}}\right) = \frac{1.1775918}{A^{2/3}}, \tag{3.2.84}$$

where $a_1 = 1.159595266\ldots$.

Comparing these expressions for $\Omega(A)$ to the exact value, the following percentage errors are found

$$E_1^{(1)} = 13.7\% \text{ error}, \quad E_1^{(2)} = 2.0\% \text{ error}, \tag{3.2.85}$$

and

$$\Omega_{\text{exact}}(A) < \Omega_1^{(2)}(A) < \Omega_1^{(1)}(A). \tag{3.2.86}$$

The second-order harmonic balance method can only be applied to the second member of Eq. (3.2.82)

$$x(\ddot{x})^3 + 1 = 0.$$

For this case

$$\begin{cases} x_2(t) = A_1(\cos\theta + z\cos 3\theta), \\ \ddot{x}_2(t) = -\Omega_2^2 A_1(\cos\theta + 9z\cos 3\theta), \end{cases}$$

where $\theta = \Omega_2 t$ and $A_2 = zA_1$. From these expressions $x(\ddot{x})^3$ can be calculated and we find upon substituting into the differential equation the result

$$[-(\Omega_2^2)^3 A_1^4]\left[\left(\frac{f_1 + 2f_2}{2}\right)\right] + [-(\Omega_2^2)^3 A_1^4]\left[\frac{(f_1 + f_3)(1+z)}{2}\right]\cos 2\theta + 1 + \text{HOH} \simeq 0, \tag{3.2.87}$$

where

$$f_1 = \left(\frac{3}{4}\right) + \left(\frac{513}{2}\right)z, \quad f_2 = \left(\frac{1}{4}\right) + \left(\frac{27}{4}\right)z + \left(\frac{2187}{4}\right)z^3$$

$$f_3 = \left(\frac{135}{2}\right)z.$$

Harmonic balancing gives

$$-(\Omega_2^2)^3 A_1^4\left(\frac{f_1 + 2f_2}{2}\right) + 1 = 0, \tag{3.2.88}$$

$$-(\Omega_2^2)^3 A_1^4\left[\frac{(f_1 + 2f_2)(1+z)}{2}\right] = 0. \tag{3.2.89}$$

The second equation allows the determination of z since it can be written as

$$(f_1 + f_2)(1+z) = \left[\left(\frac{3}{4}\right) + \left(\frac{513}{2}\right)z\right](1+z) = 0,$$

and the smallest magnitude root is

$$z = -\left(\frac{3}{1026}\right) = -0.00292397\ldots. \tag{3.2.90}$$

We can now solve Eq. (3.2.88) for Ω_2 to obtain

$$\Omega_2^2 = \left[\frac{2}{\left(\frac{5}{4}\right) + \left(\frac{621}{4}\right) z + \left(\frac{2187}{2}\right) z^3}\right] \frac{1}{A_1^4},$$

and

$$\Omega_2(A_1) = \frac{1.190568}{A_1^{2/3}}. \tag{3.2.91}$$

Since $A_1 = A/(1+z)$, we have

$$\Omega_1(A) = \left[\frac{1.190568}{A^{2/3}}\right] (1+z)^{2/3} = \frac{1.188246}{A^{2/3}}. \tag{3.2.92}$$

Therefore, in comparison to $\Omega_{\text{exact}}(A)$, the percentage error is

$$E_2 = 2.9\%. \tag{3.2.93}$$

From Eqs. (3.2.85) and (3.2.93), we learn that the percentage error for the angular frequency is slightly larger for the second-order harmonic balance approximation in comparison to a first-order calculation.

In summary, the second-order method of harmonic balance gives the following answer for the periodic solution of the ICR oscillator

$$x_2(t) = \left(\frac{1026}{1023}\right) A \left\{\cos[\Omega_2(A)t] - \left(\frac{3}{1026}\right) \cos[3\Omega_2(A)t]\right\}. \tag{3.2.94}$$

3.3 Rational Approximations [23, 24]

A useful alternative procedure for calculating a second-order harmonic balance approximation is the rational approximation. This technique was introduced by Mickens [23] and has been extended in its applications by Beléndez et al. [2]. A major advantage of the rational approximation is that it gives an implicit inclusion of all the harmonics contributing to the periodic solutions. This rational form is given by the expression

$$x(t) = \frac{A_1 \cos\theta}{1 + B_1 \cos 2\theta}, \quad \theta = \Omega_1 t \tag{3.3.1}$$

where (A_1, B_1, Ω_1) are, for the moment, unknown constants. For a particular application, they are determined as functions of the initial conditions and the mathematical structure of the oscillatory differential equation.

In this section, we enumerate several of the properties associated with Eq. (3.3.1).

3.3.1 *Fourier Expansion* [24]

The right-hand side of Eq. (3.3.1) is periodic in the variable θ with period 2π. Further, it has the Fourier representation [23]

$$x(t) = \sum_{k=1}^{\infty} a_k \cos(2k+1)\theta, \tag{3.3.2}$$

where the coefficients may be calculated from the expression [9]

$$a_k = \frac{1}{\pi}\int_{-\pi}^{\pi}\left[\frac{A_1\cos\theta}{1+B_1\cos 2\theta}\right]\cos(2k+1)\theta\, d\theta. \tag{3.3.3}$$

The integral can be evaluated using

$$\int_0^{\pi}\frac{\cos(n\phi)d\phi}{1+b\cos\phi} = \frac{\pi(-1)^n}{\sqrt{1-b^2}}\left[\frac{1-\sqrt{1-b^2}}{b}\right]^n,$$

provided $|b| < 1$; see [16, p. 366, Eq. (3.613.1)]. Using this result, it may be shown that the a_k are given by the formula [24]

$$a_k = \frac{(-1)^k A}{\sqrt{1-B_1^2}}\left[\frac{1-\sqrt{1-B_1^2}}{B_1}\right]^k\left[\frac{B_1 - 1 + \sqrt{1-B_1^2}}{B_1}\right], \tag{3.3.4}$$

for

$$|B_1| < 1. \tag{3.3.5}$$

Note that the result in Eq. (3.3.1) is not meaningful unless the condition of Eq. (3.3.5) holds. Inspection of Eq. (3.3.4) shows that it provides a full characterization of the Fourier coefficients of the rational harmonic balance representation as given in Eq. (3.3.1).

3.3.2 *Properties of a_k*

If we assume that

$$|B_1| \ll 1, \tag{3.3.6}$$

then Eq. (3.3.4) can be written

$$a_k = \begin{cases} (-1)^k A e^{-ak}, & 0 < B \ll 1, \\ A e^{-ak}, & 0 < (-B) \ll 1, \end{cases} \tag{3.3.7}$$

where "a" is

$$e^{-a} = \frac{|B_1|}{2}. \tag{3.3.8}$$

This last result implies that the Fourier coefficients for the rational harmonic balance approximation decrease exponentially in the index k.

In summary, Eq. (3.3.1) provides an approximation to all of the harmonics for the exact solution, and the coefficients decrease exponentially [23, 24].

3.3.3 *Calculation of $\ddot{x}$*

In calculations involving the rational harmonic balance approximation, the second derivative is required to be evaluated. Starting with

$$x(t) = \frac{A_1 \cos\theta}{1 + B_1 \cos 2\theta}, \quad \theta = \Omega_1 t,$$

it follows that $\ddot{x}$ is

$$(1 + B_1 \cos 2\theta)^3 \ddot{x} = -(\Omega_1^2 A_1)\left\{\left[1 + B_1 - \left(\frac{11}{2}\right) B_1^2\right] \cos\theta + 3B_1 \left[\frac{3B_1}{4} - 1\right] \cos 3\theta + \text{HOH}\right\}. \quad (3.3.9)$$

This formula will be of value for the calculations to be completed in the next section.

3.4 Worked Examples

To illustrate the utility of the rational harmonic balance formulation, this section contains details of the calculations for three TNL oscillator differential equations. This task, in each case, is to determine A_1, B_1, and Ω_1, in terms of the initial conditions.

3.4.1 $\ddot{x} + x^3 = 0$

We begin by observing that

$$(1 + B_1 \cos 2\theta)^3 x^3 = (A_1 \cos\theta)^3 = \left(\frac{3A_1^3}{4}\right) \cos\theta + \left(\frac{A_1^3}{4}\right) \cos 3\theta. \quad (3.4.1)$$

Using this result and Eq. (3.3.9), it follows that

$$\ddot{x} + x^3 = 0 \quad (3.4.2)$$

can be written as

$$\left\{-(\Omega_1^2 A_1)\left[1 + B_1 - \left(\frac{11}{2}\right) B_1^2\right] + \frac{3A_1^3}{4}\right\} \cos\theta + \left\{-(\Omega_1^2 A_1)(3B_1)\left[\frac{3B_1}{4} - 1\right] + \frac{A_1^3}{4}\right\} \cos 3\theta + \text{HOH}. \quad (3.4.3)$$

Setting the coefficients of $\cos\theta$ and $\cos 3\theta$ to zero gives

$$\Omega_1^2 \left[1 + B_1 - \left(\frac{11}{2}\right) B_1^2\right] = \left(\frac{3}{4}\right) A_1^2, \quad (3.4.4)$$

$$\Omega_1^2(3B_1)\left[\frac{3B_1}{4}-1\right]=\frac{A_1^2}{4}. \tag{3.4.5}$$

A separate equation involving only B_1 can be found by dividing these equations; doing so gives the result

$$\frac{1+B_1-\left(\frac{11}{2}\right)B_1^2}{3B_1\left[\frac{3B_1}{4}-1\right]}=3,$$

or

$$\left(\frac{49}{4}\right)B_1^2-10B_1-1=0, \tag{3.4.6}$$

and the root having the smallest magnitude is

$$B_1=-0.090064. \tag{3.4.7}$$

If Eq. (3.4.4) is solved for Ω_1^2, then

$$\Omega_1^2=\left(\frac{3}{4}\right)A_1^2\Big/\left[1+B_1-\left(\frac{11}{2}\right)B_1^2\right]=(0.866728)A_1^2 \tag{3.4.8}$$

or

$$\Omega_1=(0.930982)A_1.$$

However, for $x(0)=A$ and $\dot{x}(0)=0$, it follows that

$$A=\frac{A_1}{1+B_1}$$

or

$$A_1=(1+B_1)A.$$

Therefore,

$$\Omega_1(A)=(0.930982)(1+B_1)A=(0.847134)A, \tag{3.4.9}$$

and the corresponding period is

$$T_1(A)=\frac{2\pi}{\Omega_1(A)}=\frac{7.4170}{A}. \tag{3.4.10}$$

Since the exact period is

$$T_{\text{exact}}(A)=\frac{7.4163}{A},$$

the percentage error is

$$\left|\frac{T_{\text{exact}}-T_1}{T_{\text{exact}}}\right|\cdot 100=0.01\%\ \text{error}. \tag{3.4.11}$$

This calculation indicates that the rational harmonic balance representations give excellent estimates for the angular frequency and period. Therefore, we have for this approximation the result

$$x(t)=\frac{(0.909936)A\cos[(0.847134)At]}{1-(0.090064)\cos[(1.694268)At]}. \tag{3.4.12}$$

3.4.2 $\ddot{x} + x^2 sgn(x) = 0$ [25]

The quadratic TNL oscillator can be written in either of the forms

$$\begin{cases} \ddot{x} + x^2 \text{sgn}(x) = 0, \\ \ddot{x} + |x|x = 0, \end{cases} \tag{3.4.13}$$

with initial conditions, $x(0) = A$, and $\dot{x}(0) = 0$, holding for each equation. With the rational representation

$$x(t) = \frac{A_1 \cos\theta}{1 + B_1 \cos 2\theta}, \quad \theta = \Omega_1 t,$$

then

$$|x(t)| = \frac{A_1 |\cos\theta|}{1 + B_1 \cos 2\theta}, \tag{3.4.14}$$

and we need to find an expression for the Fourier expansion of $|\cos\theta|$. An easy and direct calculation gives

$$|\cos\theta| = \left(\frac{4}{\pi}\right) \left[\left(\frac{1}{2}\right) + \left(\frac{1}{3}\right) \cos 2\theta - \left(\frac{1}{15}\right) \cos 4\theta + \cdots \right], \tag{3.4.15}$$

and

$$\begin{aligned} (1 + B_1 \cos 2\theta)^3 |x|x = \left(\frac{8A_1^2}{3\pi}\right) \Bigg\{ & \left[1 + \frac{3B_1}{5}\right] \cos\theta \\ & + \left(\frac{1}{5}\right) \left[1 + \frac{17B_1}{7}\right] \cos 3\theta + \text{HOH} \Bigg\}. \end{aligned} \tag{3.4.16}$$

Substituting Eqs. (3.3.9) and (3.4.16) into the second of Eqs. (3.4.13), and equating to zero, the coefficients of the resulting expressions in $\cos\theta$ and $\cos 3\theta$, gives

$$\left(\frac{8A_1^2}{3\pi}\right) \left[1 + \frac{3B_1}{5}\right] = \Omega_1^2 A_1 \left[1 + B_1 - \left(\frac{11}{2}\right) B_1^2\right], \tag{3.4.17}$$

$$\left(\frac{8A_1^2}{15\pi}\right) \left[1 + \left(\frac{17}{7}\right) B_1\right] = \Omega_1^2 A_1 \left[\left(\frac{9}{4}\right) B_1^2 - 3B_1\right]. \tag{3.4.18}$$

To obtain an equation for B_1, divide these two equations to obtain, after simplification, the expression

$$\left(\frac{563}{28}\right) B_1^3 + \left(\frac{169}{14}\right) B_1^2 - \left(\frac{129}{7}\right) B_1 - 1 = 0, \tag{3.4.19}$$

which has as its smallest magnitude root

$$B_1 = -0.052609. \tag{3.4.20}$$

With this B_1, Ω_1^2 can be found from Eq. (3.4.18)

$$\Omega_1^2 = \left(\frac{8}{15\pi}\right)\left[\frac{1+\left(\frac{17}{7}\right)B_1}{\left(\frac{9}{4}\right)B_1^2 - 3B_1}\right]A_1. \tag{3.4.21}$$

Since $A_1 = (1+B_1)A$, we have

$$\Omega_1^2(A) = \left(\frac{8}{15\pi}\right)\left[\frac{1+\left(\frac{17}{7}\right)B_1}{\left(\frac{9}{4}\right)B_1^2 - 3B_1}\right]\left(\frac{A}{1+B_1}\right), \tag{3.4.22}$$

and

$$\Omega_1(A) = (0.95272)A^{1/2}. \tag{3.4.23}$$

Substituting these quantities into $x(t)$, we obtain the following expression

$$x(t) = \frac{(0.9474)A\cos[(0.9527)A^{1/2}t]}{1-(0.0526)\cos[(1.9054)A^{1/2}t]}. \tag{3.4.24}$$

3.4.3 $\ddot{x} + x^{-1} = 0$

This TNL oscillator equation can be written as

$$x\ddot{x} + 1 = 0. \tag{3.4.25}$$

Substituting the rational approximations for x and $\ddot{x}$ gives

$$\begin{aligned} -(\Omega_1^2 A_1)\left[\frac{A_1\cos\theta}{(1+B_1\cos 2\theta)^4}\right]&\left\{\left[1+B_1-\left(\frac{11}{2}\right)B_1^2\right]\cos\theta\right. \\ &\left. + 3B_1\left[\frac{3B_1}{4}-1\right]\cos 3\theta + \text{HOH}\right\} + 1 \simeq 0, \end{aligned} \tag{3.4.26}$$

$$\begin{aligned} -(\Omega_1^2 A_1^2)(\cos\theta)&\left\{\left[1+B_1-\left(\frac{11}{2}\right)B_1^2\right]\cos\theta\right. \\ &\left. + 3B_1\left[\frac{3B_1}{4}-1\right]\cos 3\theta + \text{HOH}\right\} \\ &+ (1+B_1\cos 2\theta)^4 + \text{HOH} \simeq 0. \end{aligned} \tag{3.4.27}$$

Expanding $(\cos\theta)^2$ and $\cos\theta\cos 3\theta$ in the big-bracketed expressions gives

$$-(\Omega_1^2 A_1^2)\left\{\left(\frac{1}{2}\right)\left[1+B_1-\left(\frac{11}{2}\right)B_1^2\right]\right.$$

$$+\left(\frac{1}{2}\right)\left[1-2B_1-\left(\frac{13}{4}\right)B_1^2\right]\cos 2\theta + \text{HOH}\Big\}.$$

Likewise, expanding $(1+B_1\cos 2\theta)^4$ gives

$$(1+B_1\cos 2\theta)^4 = \left[1+3B_1^2+\left(\frac{3}{8}\right)B_1^4\right] + [B_1(4+3B_1^2)]\cos 2\theta + \text{HOH}.$$

Setting the coefficients of the constant and $\cos 2\theta$ to zero gives

$$\Omega_1^2 A_1^2\left(\frac{1}{2}\right)\left[1+B_1-\left(\frac{11}{2}\right)B_1^2\right] = 1+3B_1^2+\left(\frac{3}{8}\right)B_1^4, \tag{3.4.28}$$

$$\Omega_1^2 A_1^2\left(\frac{1}{2}\right)\left[1-2B_1-\left(\frac{13}{4}\right)B_1^2\right] = B_1+(4+3B_1^2). \tag{3.4.29}$$

Dividing the two equations gives

$$\left(\frac{39}{32}\right)B_1^6-\left(\frac{63}{4}\right)B_1^5+\left(\frac{51}{4}\right)B_1^4-(13)B_1^3 + \left(\frac{17}{4}\right)B_1^2+6B_1-1=0, \tag{3.4.30}$$

and the smallest, in magnitude, root is

$$B_1 = 0.15662. \tag{3.4.31}$$

If Eq. (3.4.28) is solved for Ω_1^2 and A_1 is replaced by $A(1+B_1)$, then

$$\Omega_1^2 = 2\left[\frac{1+3B_1^2+\left(\frac{3}{8}\right)B_1^4}{1+B_1-\left(\frac{11}{2}\right)B_1^2}\right]\frac{1}{(1+B_1)^2A^2}$$

and therefore

$$\Omega_1(A) = \frac{1.25350}{A}. \tag{3.4.32}$$

The percentage error, in comparison with the exact solution [14]

$$\Omega_{\text{exact}}(A) = \frac{1.25331}{A}$$

is

$$\left|\frac{\Omega_{\text{exact}}-\Omega_1}{\Omega_{\text{exact}}}\right|\cdot 100 = 0.015\%\ \text{error}. \tag{3.4.33}$$

3.5 Third-Order Equations

Nonlinear, third-order differential equations may be used to model various physical phenomena. Particular cases include stellar oscillations [6, 26], and third-order mechanical oscillators [27, 28]. This section gives a brief discussion of the Castor model [6] and several related TNL generalizations.

3.5.1 *Castor Model*

The Castor Model is

$$\ddot{x} + \dot{x} + \epsilon(x - x^3) = 0. \tag{3.5.1}$$

The corresponding three system equations are

$$\frac{dx}{dy} = 0, \quad \frac{dy}{dt} = z, \quad \frac{dz}{dt} = -y - \epsilon(x - x^3) = 0; \tag{3.5.2}$$

they describe the motion along trajectories in a three-dimensional phase-space (x, y, z). This system has three fixed points or equilibrium solutions

$$\begin{aligned} (\bar{x}^{(1)}, \bar{y}^{(1)}, \bar{z}^{(1)}) &= (-1, 0, 0), \\ (\bar{x}^{(2)}, \bar{y}^{(2)}, \bar{z}^{(2)}) &= (0, 0, 0), \\ (\bar{x}^{(3)}, \bar{y}^{(3)}, \bar{z}^{(3)}) &= (+1, 0, 0). \end{aligned} \tag{3.5.3}$$

However, only the second fixed point is physically relevant to the understanding of stellar oscillations [26].

Assume that a first-order harmonic balance procedure may be applied to the Castor model, i.e., we take the approximation to the periodic solution to be

$$x(t) \simeq A\cos(\Omega t), \tag{3.5.4}$$

where, for the moment, A and Ω are unknown parameters. Substitution of this expression into Eq. (3.5.1) gives, for $\theta = \Omega t$, the result

$$(\Omega^3 A \sin\theta) + (-\Omega A \sin\theta) + \epsilon[A\cos\theta - (A\cos\theta)^3] \simeq 0$$

and

$$A\Omega(\Omega^2 - 1)\sin\theta + \epsilon A\left[1 - \left(\frac{3}{4}\right)A^2\right]\cos\theta + \text{HOH} \simeq 0. \tag{3.5.5}$$

Setting the coefficients of $\sin\theta$ and $\cos\theta$ to zero gives

$$A\Omega(\Omega^2 - 1) = 0, \quad A\left[1 - \left(\frac{3}{4}\right)A^2\right] = 0. \tag{3.5.6}$$

If $A = 0$, then $x(t) = 0$ and this is the equilibrium solution $(\bar{x}^{(2)}, \bar{y}^{(2)}, \bar{z}^{(2)}) = (0, 0, 0)$. For $A \neq 0$, it follows that

$$\Omega = 1, \quad A = \sqrt{\frac{4}{3}} \tag{3.5.7}$$

and

$$x(t) \simeq \sqrt{\frac{4}{3}}\cos t. \tag{3.5.8}$$

Observe that this has a definite amplitude and angular frequency. It is a limit-cycle [1, 6, 28] solution for the Castor model.

3.5.2 *TNL Castor Models*

The following four equations are generalizations of the Castor model to include TNL functions

$$\dddot{x} + \dot{x} + \epsilon(x^{1/3} - x^3) = 0, \tag{3.5.9}$$

$$\dddot{x} + \dot{x} + \epsilon(x^{1/3} - x) = 0, \tag{3.5.10}$$

$$\dddot{x} + \dot{x}^{1/3} + \epsilon(x^{1/3} - x) = 0, \tag{3.5.11}$$

$$\dddot{x} + (\dot{x})^2\text{sgn}(x) + \epsilon(x^{1/3} - x) = 0. \tag{3.5.12}$$

We now construct a first-order harmonic balance approximation to the solution of Eq. (3.5.11). Starting with

$$x(t) \simeq A\cos\theta, \quad \theta = \Omega t,$$

we obtain upon substitution the result

$$\Omega^3 A\sin\theta + (-\Omega A\sin\theta)^{1/3} + \epsilon\left[(A\cos\theta)^{1/3} - A\cos\theta\right] \simeq 0. \tag{3.5.13}$$

Using

$$\begin{cases} (\cos\theta)^{1/3} = a_1\cos\theta + \text{HOH}, \\ (\sin\theta)^{1/3} = a_1\sin\theta + \text{HOH}, \\ a_1 = 1.159595\ldots, \end{cases} \tag{3.5.14}$$

see Mickens [9, Section 2.7], we find for Eq. (3.5.13) the result

$$[\Omega^3 A - \Omega^{1/3}A^{1/3}a_1]\sin\theta + \epsilon[A^{1/3}a_1 - A]\cos\theta + \text{HOH} \simeq 0,$$

which can be rewritten to the form

$$(\Omega^{1/3}A^{1/3})[\Omega^{8/3}A^{2/3} - a_1]\sin\theta + \epsilon A^{1/3}[a_1 - A^{2/3}] + \text{HOH} \simeq 0. \tag{3.5.15}$$

Harmonic balancing this expression gives

$$\Omega^{1/3}A^{1/3}[\Omega^{8/3}A^{2/3} - a_1] = 0,$$

$$A^{1/3}[a_1 - A^{2/3}] = 0,$$

with the nontrivial solution

$$A = (a_1)^{3/2}, \quad \Omega = 1. \tag{3.5.16}$$

Therefore, Eq. (3.5.11) has a limit-cycle and the approximation to this periodic solution is

$$x(t) \simeq (a_1)^{3/2}\cos t. \tag{3.5.17}$$

3.6 Resume

We end this chapter by listing several advantages and disadvantages of the harmonic balance method, especially as it relates to the calculating approximations to the periodic solutions of TNL oscillator differential equations.

3.6.1 *Advantages*

- Harmonic balancing can be applied to both standard and TNL oscillator equations.
- It is easy and straightforward to formulate the functional forms for approximating the periodic solutions.
- For certain equations a first-order calculation can provide very "accurate" results, especially as measured in terms of the percentage error for the angular frequency, Ω.
- Generally, both first- and second-order harmonic balance approximations can be done by hand, i.e., the associated mathematical work does not require the use of packaged software.
- Standard harmonic balancing methods are based on trigonometric functions. However, it is possible to formulate these procedures using any complete set of periodic functions; examples of such functions are the Jacobi elliptic functions [29–31].

3.6.2 *Disadvantages*

- TNL oscillator differential equations containing terms raised to a fractional power or terms that have discontinuities may have to be rewritten to a form suitable for the application of harmonic balancing methods. Currently, no *a priori* procedures exist for determining which modified equation structure to use for a particular TNL equation.
- Calculating the amplitudes and the angular frequency may become algebraically intensive.
- It is *a priori* difficult to predict for a given TNL equation whether a first-order harmonic balance calculation will provide a sufficiently accurate approximation to the periodic solutions.
- Existing formulations of the harmonic balance procedure do not allow it to be applied to non-conservative oscillators. These types of oscillators have solutions involving transient behaviors [1, 11, 12, 28].

Problems

3.1 Construct a third-order harmonic balance approximation for the TNL oscillator equation

$$\ddot{x} + x^3 = 0.$$

3.2 Equation (3.2.11) gives the cubic equation

$$51z^3 + 27z^2 + 21z - 1 = 0.$$

Give reasons to justify why $z \approx 1/21$ is expected to be close in value to the smallest magnitude solution of the cubic equation. Can you generalize this result? See also Eqs. (3.2.47) and (3.2.48).

3.3 Provide reasons why the percentage error is a better gauge of the "error" rather than the absolute error, i.e., $|$(exact value) $-$ (calculated value)$|$.

3.4 Evaluate the integral appearing in Eq. (3.2.22), i.e.,

$$\int_0^1 \frac{ds}{\sqrt{\ln\left(\frac{1}{s}\right)}}.$$

3.5 Are the two differential equations

$$\ddot{x} + x^{1/3} = 0, \qquad (\dot{x})^3 + x = 0,$$

mathematically equivalent? Explain your answer.

3.6 Calculate the Fourier series for $|\cos\theta|$ and use this result to obtain the Fourier series of $|\sin\theta|$.

3.7 Determine the Fourier expansion for

$$\frac{1}{\cos\theta}$$

and use this result to calculate a first-order harmonic balance approximation to the periodic solution of

$$\ddot{x} + \frac{1}{x} = 0.$$

3.8 Derive the expression for ϕ, Eq. (3.2.58), from Eq. (3.2.56).

3.9 Show that the expression, given in Eq. (3.2.75), is a closed curve in the (x, y) phase-space.

3.10 The rational harmonic balance discussed in Section 3.3 contains only cosine terms. Are there circumstances in which the form

$$x(t) = \frac{A_1 \cos\theta}{1 + B_1 \sin 2\theta}$$

might be useful?

3.11 In Eq. (3.3.1) and in Problem 3.10, what are the physical and mathematical reasons for expecting $|B_1| < 1$? What happens if this condition does not hold?
3.12 Derive Eq. (3.3.4).
3.13 Derive the results given by the expression in Eq. (3.3.9).
3.14 Explain why the result for $|x(t)|$ is correct in Eq. (3.4.14).
3.15 Construct a rational harmonic balance approximation to the periodic solution of

$$\ddot{x} + x^{1/3} = 0.$$

3.16 Calculate first-order harmonic balance solutions for Eqs. (3.5.9), (3.5.10), and (3.5.12).

References

[1] R. E. Mickens, *Oscillations in Planar Dynamic Systems* (World Scientific, Singapore, 1996).
[2] A. Beléndez, E. Gimeno, M. L. Álvarez, S. Gallego, M. Ortunõ, and D. I. Méndez, *Journal of Nonlinear Sciences and Numerical Simulation* **10**, 13 (2009).
[3] C. A. Borges, L. Cesari, and D. A. Sanchez, *Quarterly of Applied Mathematics* **32**, 457 (1975).
[4] P. Miletta, in R. Chuagui (editor), *Analysis, Geometry and Probability* (Marcel Dekker, New York, 1985). See pages 1–12.
[5] N. A. Bobylev, Y. M. Burman, and S. K. Korovin, *Approximation Procedures in Nonlinear Oscillation Theory* (Walter deGruyter, Berlin, 1994). See Chapter 1, Section 4.
[6] J. P. Cox, *Theory of Stellar Pulsation* (Princeton University Press; Princeton, NJ; 1980).
[7] R. E. Mickens, *Journal of Sound and Vibration* **94**, 456 (1984).
[8] R. E. Mickens, *Journal of Sound and Vibration* **118**, 561 (1987).
[9] R. E. Mickens, *Mathematical Methods for the Natural and Engineering Sciences* (World Scientific, Singapore, 2004).
[10] R. E. Mickens, *Journal of Sound and Vibration* **258**, 398 (2000).
[11] A. H. Nayfeh, *Perturbation Methods* (Wiley, New York, 1973).
[12] N. Minorsky, *Nonlinear Oscillations* (Kreiger; Huntington, NY; 1974).
[13] See ref. [1], Sections 4.3.1 and 4.5.1.
[14] R. E. Mickens, *Journal of Sound and Vibration* **306**, 968 (2007).
[15] J. R. Acton and P. T. Squire, *Solving Equations with Physical Understanding* (Adam Hilger, Boston, 1985). See Chapter 5.
[16] I. S. Gradshteyn and I. M. Ryzhik, *Tables of Integrals, Series and Products* (Academic Press, New York, 1980).

[17] R. E. Mickens, "Exact solutions for the quadratic oscillator," Clark Atlanta University; Atlanta, GA; June 15, 2009 (unpublished results).
[18] R. E. Mickens, *Journal of Sound and Vibration* **159**, 546 (1992).
[19] R. E. Mickens, *Journal of Sound and Vibration* **246**, 375 (2001).
[20] R. E. Mickens, *Journal of Sound and Vibration* **255**, 789 (2002).
[21] R. E. Mickens, *Journal of Sound and Vibration* **292**, 964 (2006).
[22] The work presented in this section was done jointly with Mr. Dorian Wilkerson and forms the basis of the doctoral dissertation for the Ph.D. Degree in System Sciences at Clark Atlanta University (2009).
[23] R. E. Mickens, *Journal of Sound and Vibration* **111**, 515 (1986).
[24] R. E. Mickens and D. Semwogere, *Journal of Sound and Vibration* **195**, 528 (1996).
[25] See Mickens [1], Section 4.5.2.
[26] R. E. Mickens, *Computers and Mathematics with Applications* **57**, 740 (2009).
[27] H. P. W. Gottlieb, *Journal of Sound and Vibration* **271**, 671 (2004).
[28] Y. A. Mitropolskii and N. V. Dao, *Applied Asymptotic Methods in Nonlinear Oscillations* (Kluwer Academic Publishers, Dordrecht, 1997).
[29] J. Garcia-Margallo and J. Diaz Bejarano, *Journal of Sound and Vibration* **116**, 591 (1987).
[30] S. Bravo Yuste, *Journal of Sound and Vibration* **130**, 33 (1989); **45**, 381 (1991).
[31] J. Garcia-Margallo and J. Diaz Bejarano, *Journal of Sound and Vibration* **136**, 453 (1990).

Chapter 4

Parameter Expansions

4.1 Introduction

The parameter "insertion" and "expansion" methodology was introduced in a paper by Senator and Bapat [1]. Subsequently, it was extended in a publication of Mickens [2]. However, the full generalization of this concept was done by He [3]. Under his direction this method has been applied to a variety of equations in many areas where nonlinear differential equations model interesting and important physical and engineering phenomena. The excellent review paper by He [4] gives a broad overview of this technique.

In broad terms, the parameter expansion methodology consists of the following steps:

(1) First, a parameter p is introduced, where

$$0 \leq p \leq 1, \tag{4.1.1}$$

and the original differential equation is rewritten to a form such that for $p = 1$, the original equation is recovered.

(2) Second, the dependent variable and one or more of its "constants" are expanded in a series involving powers of the parameter and the rewritten equation is then solved for $0 < p \ll 1$ using standard perturbation methods. This approximation to the solution of the rewritten equation will be denoted by $x(p, t)$.

(3) Finally, the function $x(p, t)$ is evaluated at $p = 1$ and this is taken as a valid approximation for the solution of the original equation.

In principle, the parameter expansion procedure may be applied to any class of mathematical equations, although, in practice, its use has been limited to nonlinear differential equations. In the following section we demonstrate by means of several worked examples the actual application of this technique.

4.2 Worked Examples

In the following five examples, we will first obtain the "p" expansion equations and solve them under the assumption that p is small, i.e., $0 < p \ll 1$. From that point on, the final value, $p = 1$, will be used for the remainder of the calculation. For all of these calculations, it will be assumed that the initial conditions are

$$x(0) = A, \quad \dot{x}(0) = 0. \tag{4.2.1}$$

4.2.1 $\ddot{x} + x^3 = 0$

This TNL, second-order differential equation

$$\ddot{x} + x^3 = 0 \tag{4.2.2}$$

contains no linear term in x. However, let us consider the following equation

$$\ddot{x} + 0 \cdot x + px^3 = 0, \tag{4.2.3}$$

and "p-expand" the constant zero [5] and the solution, i.e.,

$$0 = \Omega^2 + p\omega_1 + \cdots, \tag{4.2.4}$$

$$x = x_0 + px_1 + \cdots, \tag{4.2.5}$$

where Ω^2, ω_1, x_0, and x_1 are to be determined. Note that when $p = 1$, Eq. (4.2.3) reduces to Eq. (4.2.2). Substitution of Eqs. (4.2.4) and (4.2.5) into Eq. (4.2.3) gives

$$(\ddot{x} + p\ddot{x}_1 + \cdots) + (\Omega^2 + p\omega_1 + \cdots)(x_0 + px_1 + \cdots) + p(x_0 + px_1 + \cdots)^3 = 0. \tag{4.2.6}$$

If the terms of order p^0 and p are collected together and equated to zero, we obtain

$$p^0 : \ddot{x}_0 + \Omega^2 x_0 = 0, \quad x_0(0) = A, \quad \dot{x}_0(0) = 0; \tag{4.2.7}$$

$$p : \ddot{x}_1 + \Omega^2 x_1 = -\omega_1 x_0 - x_0^3, \quad x_1(0) = 0, \quad \dot{x}_1(0) = 0. \tag{4.2.8}$$

The solution for x_0 is

$$x_0(t) = A\cos\theta, \quad \theta = \Omega t, \tag{4.2.9}$$

and the substitution of this into Eq. (4.2.8) gives

$$\begin{aligned}\ddot{x}_1 + \Omega^2 x_1 &= \omega_1 A\cos\theta - (A\cos\theta)^3 \\ &= -A\left[\omega_1 + \frac{3A^2}{4}\right]\cos\theta - \left(\frac{A^3}{4}\right)\cos 3\theta. \end{aligned} \tag{4.2.10}$$

The elimination of secular terms requires that the following condition is satisfied

$$A\left[\omega_1 + \frac{3A^2}{4}\right] = 0 \Rightarrow \omega_1 = -\frac{3A^2}{4}.$$

To first-order in p, with p set to one, Eq. (4.2.4) gives

$$\Omega^2 = -\omega_1 = \frac{3A^2}{4}. \tag{4.2.11}$$

Therefore, $x_1(t)$ is the solution to the differential equation

$$\ddot{x}_1 + \Omega^2 x_1 = -\left(\frac{A^3}{4}\right)\cos 3\theta, \tag{4.2.12}$$

where

$$x_1(0) = 0, \quad \dot{x}_1(0) = 0, \quad \theta = \Omega t = \left(\frac{3}{4}\right)^{1/2} At. \tag{4.2.13}$$

The particular solution to Eq. (4.2.12) is [6]

$$x_1^{(p)}(t) = C\cos 3\theta,$$

where, upon substitution into Eq. (4.2.12), we obtain

$$(-9\Omega^2 + \Omega^2)C = -\frac{A^3}{4}$$

or

$$C = \frac{A^3}{32\Omega^2} = \frac{A^3}{(32)\left(\frac{3A^2}{4}\right)} = \frac{A}{24}. \tag{4.2.14}$$

Thus, the general solution for $x_1(t)$ is

$$x_1(t) = D\cos\theta + \left(\frac{A}{24}\right)\cos 3\theta$$

and with $x_1(0) = 0$, it follows that $D = -A/24$ and

$$x_1(t) = -\left(\frac{A}{24}\right)(\cos\theta - \cos 3\theta). \tag{4.2.15}$$

Therefore, the p-parameter solution, to order-one, with $p = 1$, for Eq. (4.2.2) is

$$x(t) = x_0(t) + x_1(t) = A\cos\theta - \left(\frac{A}{24}\right)(\cos\theta - \cos 3\theta)$$

or

$$x(t) = A\left\{\left(\frac{23}{24}\right)\cos\left[\left(\frac{3}{4}\right)^{1/2} At\right] + \left(\frac{1}{24}\right)\cos\left[3\left(\frac{3}{4}\right)^{1/2} At\right]\right\}. \tag{4.2.16}$$

4.2.2 $\ddot{x} + x^{-1} = 0$

This TNL oscillator equation

$$\ddot{x} + \frac{1}{x} = 0, \tag{4.2.17}$$

can be rewritten to the form

$$x^2\ddot{x} + x = 0. \tag{4.2.18}$$

Within the p-expansion formulation, we start with the expression [4]

$$0 \cdot \ddot{x} + 1 \cdot x + px^2\ddot{x} = 0, \tag{4.2.19}$$

where to first-order in p

$$\begin{cases} 0 = 1 + pb_1 + \cdots, \\ 1 = \Omega^2 + pa_1 + \cdots, \\ x = x_0 + px_1 + \cdots. \end{cases} \tag{4.2.20}$$

Substituting the terms of Eq. (4.2.20) into Eq. (4.2.19), collecting together the quantities of powers p^0 and p, and setting them to zero, gives

$$\ddot{x}_0 + \Omega^2 x_0 = 0, \quad x_0(0) = A, \quad \dot{x}_0(0) = 0, \tag{4.2.21}$$

$$\ddot{x}_1 + \Omega^2 x_1 + b_1\ddot{x}_0 + a_1 x_0 + \ddot{x}_0 x_0^2 = 0, \quad x_1(0) = 0, \ \dot{x}_1(0) = 0. \tag{4.2.22}$$

Absence of secular terms in the solution for $x_1(t)$ requires that

$$-b_1\Omega^2 + a_1 - \left(\frac{3}{4}\right) A^2\Omega^2 = 0. \tag{4.2.23}$$

Note that we used $x_0(t) = A\cos\theta$, $\theta = \Omega t$, for the appropriate terms in Eq. (4.2.22). To first-order in p, with $p = 1$, we have, from Eq. (4.2.20)

$$\Omega^2 + a_1 = 1, \quad b_1 = -1. \tag{4.2.24}$$

Therefore,

$$\Omega^2 = \left(\frac{4}{3}\right)\left(\frac{1}{A^2}\right),$$

or

$$\Omega = \left(\frac{4}{3}\right)^{1/2}\left(\frac{1}{A}\right) = \frac{1.1547}{A}, \tag{4.2.25}$$

a value with a percentage error of 7.9% in comparison with the exact value

$$\Omega_{\text{exact}} = \left(\frac{\pi}{2}\right)^{1/2}\frac{1}{A} = \frac{1.2533}{A}.$$

The full solution for $x_1(t)$ is

$$x_1(t) = \left(\frac{A^3}{32}\right)(\cos\theta - \cos 3\theta), \tag{4.2.26}$$

and $x(t)$ is, for $p = 1$,

$$\begin{aligned} x(t) &= x_0(t) + x_1(t) \\ &= A\cos\theta + \left(\frac{A^3}{32}\right)(\cos\theta - \cos 3\theta) \\ &= A\left(1 + \frac{A^2}{32}\right)\cos\theta - \left(\frac{A^3}{32}\right)\cos 3\theta, \end{aligned} \tag{4.2.27}$$

with

$$\theta = \Omega t = \left(\frac{4}{3}\right)^{1/2}\left(\frac{t}{A}\right). \tag{4.2.28}$$

An alternative form for the starting equation is

$$\ddot{x} + 0 \cdot x + p(\ddot{x})^2 = 0, \tag{4.2.29}$$

with

$$\begin{cases} x = x_0 + px_1 + \cdots, \\ 0 = \Omega^2 + pa_1 + \cdots. \end{cases} \tag{4.2.30}$$

For this formulation, we obtain

$$\ddot{x}_0 + \Omega^2 x_0 = 0, \quad x_0(0) = A, \quad \dot{x}_0(0) = 0; \tag{4.2.31}$$

$$\ddot{x}_1 = \Omega^2 x_1 = \left[a_1 + \left(\frac{3A^2}{4}\right)\Omega^4\right] A\cos\theta + \left(\frac{A^3}{4}\right)\Omega^4\cos 3\theta, \tag{4.2.32}$$

where in the expression for $x_1(t)$, we used $x_0 = A\cos(\Omega t)$. The absence of secular terms in the solution for $x_1(t)$ requires

$$a_1 + \left(\frac{3A^2}{4}\right)\Omega^4 = 0. \tag{4.2.33}$$

To first-order in the p-expansion, it follows from Eq. (4.2.30) that

$$a_1 = -\Omega^2. \tag{4.2.34}$$

Combining the latter two equations gives

$$\Omega^2 = \left(\frac{4}{3}\right)\frac{1}{A^2}, \tag{4.2.35}$$

which is the same as that given in Eq. (4.2.25). The resulting equation for $x_1(t)$

$$\ddot{x}_1 + \Omega^2 x_1 = \left(\frac{A^3}{4}\right)\Omega^4 \cos 3\theta, \tag{4.2.36}$$

with $x_1(0) = 0$ and $\dot{x}_1(0) = 0$, can now be solved and the following result is obtained

$$x_1(t) = \left(\frac{A}{24}\right)(\cos\theta - \cos 3\theta). \tag{4.2.37}$$

Therefore,

$$x(t) = x_0(t) + x_1(t) = A\left[\left(\frac{25}{24}\right)\cos\theta - \left(\frac{1}{24}\right)\cos 3\theta\right], \tag{4.2.38}$$

where $\theta = \Omega t$.

Observe that the previous solution has a different dependence of its coefficients on the value of A than the current solution, i.e., compare Eqs. (4.2.27) and (4.2.38).

4.2.3 $\ddot{x} + x^3/(1+x^2) = 0$

The Duffing-harmonic oscillator is

$$\ddot{x} + \frac{x^3}{1+x^2} = 0. \tag{4.2.39}$$

It can be rewritten to the form

$$\ddot{x} + x^2\ddot{x} + x^3 = 0,$$

$$\ddot{x} + 0\cdot x + x^2\ddot{x} + x^3 = 0. \tag{4.2.40}$$

For application of the p-expansion method, we use

$$\ddot{x} + 0\cdot x + p(x^2\ddot{x} + x^3) = 0,$$

with

$$0 = \Omega^2 + pa_1 + \cdots,$$

$$x = x_0 + px_1 + \cdots;$$

therefore,

$$(\ddot{x}_0 + p_1\ddot{x}_1 + \cdots) + (\Omega^2 + pa_1 + \cdots)x + p(x_0^2\ddot{x}_0 + x_0^3 + \cdots) = 0. \tag{4.2.41}$$

The equation for $x_1(t)$ is

$$\ddot{x}_1 + \Omega^2 x_1 = -\left[a_1 + (1-\Omega^2)\left(\frac{3A^2}{4}\right)\right]A\cos\theta$$

$$- \left(\frac{A^3}{4}\right)(1-\Omega^2)\cos 3\theta,$$

where we have used the fact that $x_0(t) = A\cos\theta$ with $\theta = \Omega t$. No secular terms give

$$a_1 + (1-\Omega^2)\left(\frac{3A^2}{4}\right) = 0, \tag{4.2.42}$$

and, for $p = 1$,

$$a_1 = -\Omega^2.$$

Solving for Ω^2, gives from Eq. (4.2.42) the result [7]

$$\Omega^2 = \frac{\left(\frac{3A^2}{A}\right)}{1+\frac{3A^2}{4}}, \tag{4.2.43}$$

and

$$\ddot{x}_1 + \Omega^2 x_1 = \left(\frac{A\Omega^2}{3}\right)\cos 3\theta. \tag{4.2.44}$$

The full solution of this equation, subject to $x_1(0) = 0$ and $\dot{x}_1(0) = 0$, is

$$x_1(t) = \left(\frac{A}{24}\right)(\cos\theta - \cos 3\theta). \tag{4.2.45}$$

Therefore, to order p, with $p = 1$, we have

$$\begin{aligned} x(t) &= x_0(t) + x_1(t) \\ &= A\left[\left(\frac{25}{24}\right)\cos\theta - \left(\frac{1}{24}\right)\cos 3\theta\right], \end{aligned} \tag{4.2.46}$$

where $\theta = \Omega t$ and Ω is obtained from Eq. (4.2.43).

4.2.4 $\ddot{x} + x^{1/3} = 0$

The cube-root oscillator equation is [9, 10]

$$\ddot{x} + x^{1/3} = 0. \tag{4.2.47}$$

This equation can be rewritten to the form

$$\ddot{x} + \Omega^2 x = \ddot{x} - \Omega^2(\ddot{x})^3,$$

where Ω^2, for the time being, is unknown. For the p-expansion, we use

$$\ddot{x} + \Omega^2 x = p[\ddot{x} - \Omega^2(\ddot{x})^3] \tag{4.2.48}$$

and expand only the solution $x(t)$, i.e.,

$$x = x_0 + px_1 + \cdots . \tag{4.2.49}$$

The coefficients of the p^0 and p terms are

$$\ddot{x}_0 + \Omega^2 x_0 = 0,$$

$$\ddot{x}_1 + \Omega^2 x_1 = \ddot{x}_0 - \Omega^2(\ddot{x}_0)^3.$$

Substituting $x_0 = A\cos\theta$, $\theta = \Omega t$, into the right-hand side of the second equation gives

$$\ddot{x}_1 + \Omega^2 x_1 = \left[-\Omega^2 + \frac{3A^2\Omega^8}{4}\right] A\cos\theta + \left(\frac{A^3\Omega^8}{4}\right)\cos 3\theta. \tag{4.2.50}$$

No secular terms in the solution for $x_1(t)$ implies that the coefficient of $\cos\theta$ is zero. From this condition it follows that

$$\Omega = \left(\frac{4}{3}\right)^{1/6} \frac{1}{A^{1/3}} = \frac{1.0491}{A^{1/3}} . \tag{4.2.51}$$

Using this result, the full solution to Eq. (4.2.50), satisfying the initial conditions, $x_1(0) = 0$ and $\dot{x}_1(0) = 0$, is

$$x_1(t) = \left(\frac{A}{24}\right)(\cos\theta - \cos 3\theta),$$

and $x(t)$ is

$$x(t) = x_0(t) + x_1(t) = A\left[\left(\frac{25}{24}\right)\cos\theta - \left(\frac{1}{24}\right)\cos 3\theta\right], \tag{4.2.52}$$

with $\theta = \Omega t$.

It is of great interest to observe that the expressions in Eqs. (4.2.46) and (4.2.52) are the same, i.e., based on the first-order p-expansion method, the approximations for the periodic solutions of the Duffing-harmonic and cube-root oscillators are given by the same function. Also, note that since the exact angular frequency is [9, 10]

$$\Omega_{\text{exact}} = \frac{1.070451}{A^{1/3}},$$

the percentage error in our calculation is 2%.

An alternative way of formulating the p-expansion is to use

$$0 \cdot \ddot{x} + 1 \cdot x = -p(\ddot{x})^3, \tag{4.2.53}$$

with

$$\begin{cases} x = x_0 + px_1 + \cdots , \\ 0 = 1 + pb_1 + \cdots , \\ 1 = \Omega^2 + pa_1 + \cdots . \end{cases} \tag{4.2.54}$$

The equation for $x_1(t)$ is

$$\begin{aligned} \ddot{x}_1 + \Omega^2 x_1 &= -b_1 \ddot{x}_0 - a_1 x_0 - (\ddot{x}_0)^3 \\ &= \left[b_1 \Omega^2 - a_1 + \frac{3A^2\Omega^6}{4} \right] \cos\theta + \left(\frac{A^3\Omega^6}{4} \right) \cos 3\theta. \end{aligned} \tag{4.2.55}$$

The elimination of a secular term in the solution for $x_1(t)$ gives

$$\Omega = \left(\frac{4}{3} \right)^{1/6} \frac{1}{A^{1/3}},$$

and

$$\ddot{x}_1 + \Omega^2 x_1 = \left(\frac{A^3\Omega^6}{4} \right) \cos\theta,$$

which has the full solution

$$x_1(t) = \left(\frac{A^3\Omega^4}{32} \right) (\cos\theta - \cos 3\theta). \tag{4.2.56}$$

Now

$$\frac{A^3\Omega^4}{32} = \left(\frac{1}{32} \right) \left(\frac{4}{3} \right)^{2/3} A^{5/3},$$

and, as a consequence,

$$\begin{aligned} x(t) &= x_0(t) + x_1(t) \\ &= A\cos\theta + \left(\frac{1}{32} \right) \left(\frac{4}{3} \right)^{2/3} A^{5/3} (\cos\theta - \cos 3\theta). \end{aligned} \tag{4.2.57}$$

Comparison of Eqs. (4.2.52) and (4.2.57) shows that they differ in their respective mathematical dependencies on A. The conclusion is that the p-expansion method does not have the quality of uniqueness of solutions for a given TNL equation.

4.2.5 $\ddot{x} + x^3 = \epsilon(1 - x^2)\dot{x}$

A TNL van der Pol type oscillator equation is

$$\ddot{x} + x^3 = \epsilon(1 - x^2)\dot{x}, \tag{4.2.58}$$

where the parameter ϵ is assumed to be small, i.e., $0 < \epsilon \ll 1$. The period and solution of a similar equation has been investigated by Mickens [11] and Özis and Yildirim [12]. We now apply the p-expansion method to obtain an approximation (to first-order in p) for its periodic solution.

To begin, we use the following form for the initiation of the calculation

$$\ddot{x} + 0 \cdot x + px^3 = p[\epsilon(1 - x^2)\dot{x}], \tag{4.2.59}$$

and make the replacements

$$x = x_0 + px_1 + \cdots,$$
$$0 = \Omega^2 + pa_1 + \cdots.$$

The equations for $x_0(t)$ and $x_1(t)$ are

$$\ddot{x}_0 + \Omega^2 x_0 = 0, \quad x_0(t) = A\cos\theta, \quad \theta = \Omega t,$$

$$\begin{aligned}\ddot{x}_1 + \Omega^2 x_1 &= -a_1 x_0 - x_0^3 + \epsilon(1 - x_0^2)\dot{x}_0 \\ &= -\left[a_1 + \frac{3A^2}{4}\right] A\cos\theta + (\epsilon A\Omega)\left[1 - \frac{A^2}{4}\right]\sin\theta \\ &\quad + \left(\frac{A^3}{4}\right)\cos 3\theta + \left(\frac{\epsilon A^3\Omega}{4}\right)\sin 3\theta. \end{aligned} \tag{4.2.60}$$

To first-order in p, with $p = 1$, we have

$$a_1 = -\Omega^2, \tag{4.2.61}$$

and the absence of secular terms, in $x_1(t)$, gives

$$a_1 + \frac{3A^2}{4} = 0, \quad 1 - \frac{A^2}{4} = 0; \tag{4.2.62}$$

therefore

$$A = 2, \quad \Omega^2 = \frac{3A^2}{4} = 3. \tag{4.2.63}$$

The full solution to the equation

$$\ddot{x}_1 + \Omega^2 x_1 = -\left(\frac{A^3}{4}\right)\cos 3\theta + \left(\frac{\epsilon\Omega A^3}{4}\right)\sin 3\theta$$

is

$$x_1(t) = -\left(\frac{A^3}{32\Omega^2}\right)(\cos\theta - \cos 3\theta) + \left(\frac{\epsilon A^3}{32\Omega}\right)(3\sin\theta - \sin 3\theta)$$

$$= -\left(\frac{1}{12}\right)(\cos\theta - \cos 3\theta) + \left(\frac{\epsilon}{4\sqrt{3}}\right)(3\sin\theta - \sin 3\theta). \quad (4.2.64)$$

Since $x(t) = x_0(t) + x_1(t)$, we have

$$x(t) = 2\cos\theta - \left(\frac{1}{12}\right)(\cos\theta - \cos 3\theta) + \left(\frac{\epsilon}{4\sqrt{3}}\right)(3\sin\theta - \sin 3\theta), \quad (4.2.65)$$

where $\theta = \Omega t = \sqrt{3}\,t$.

4.2.6 $\ddot{x} + sgn(x) = 0$

The antisymmetric constant force oscillator can be solved exactly [13]. This oscillator

$$\ddot{x} + \text{sgn}(x) = 0 \quad (4.2.66)$$

has a finite discontinuity in $f(x) = \text{sgn}(x)$ at $x = 0$. A way to resolve difficulties with this issue is to note that

$$[\text{sgn}(x)]^2 = 1,$$

and square the above equation written in the form $\ddot{x} = -\text{sgn}(x)$. Carrying out this procedure gives

$$1 - (\ddot{x})^2 = 0,$$

which on multiplication by $\ddot{x}$ becomes

$$\ddot{x} - (\ddot{x})^3 = 0.$$

For the purposes of applying the p-expansion method, we use

$$\ddot{x} + 0 \cdot x - p(\ddot{x})^3 = 0, \quad (4.2.67)$$

with

$$\begin{cases} x = x_0 + px_1 + \cdots, \\ 0 = \Omega^2 + pa_1 + \cdots. \end{cases}$$

If these expressions are substituted into Eq. (4.2.67) and the coefficients of the p^0 and p terms are set to zero, then the following equation for x_1 is obtained

$$\ddot{x}_1 + \Omega^2 x_1 = -a_1 x_0 + (\ddot{x}_0)^3, \quad x_1(0) = 0, \quad \dot{x}_1(0) = 0, \quad (4.2.68)$$

where $x_0(t) = A\cos\theta$, $\theta = \Omega t$. Substituting $x_0(t)$ into the right-hand side of Eq. (4.2.68) gives

$$\ddot{x}_1 + \Omega^2 x_1 = -\left[a_1 + \frac{3A^2\Omega^6}{4}\right] A\cos\theta - \left(\frac{A^3\Omega^6}{4}\right)\cos 3\theta. \tag{4.2.69}$$

The absence of secular terms in the solution for $x_1(t)$ gives

$$a_1 + \frac{3A^2\Omega^6}{4} = 0.$$

Therefore, for $p = 1$, a first-order in p-expansion calculation gives $a_1 = -\Omega^2$ and

$$\left(\frac{3A^2}{4}\right)\Omega^4 = 1 \Rightarrow \Omega = \left(\frac{4}{3}\right)^{1/4}\left(\frac{1}{A^{1/2}}\right). \tag{4.2.70}$$

Continuing, we find that the full solution to the $x_1(t)$ equation is

$$\begin{aligned} x_1(t) &= -\left(\frac{A^3\Omega^4}{32}\right)(\cos\theta - \cos 3\theta) \\ &= -\left(\frac{A}{24}\right)(\cos\theta - \cos 3\theta), \end{aligned} \tag{4.2.71}$$

and

$$x(t) = x_0(t) + x_1(t) = A\left[\left(\frac{23}{24}\right)\cos\theta + \left(\frac{1}{24}\right)\cos 3\theta\right], \tag{4.2.72}$$

where

$$\theta = \Omega t = \left(\frac{4}{3}\right)^{1/4}\left(\frac{t}{A^{1/2}}\right). \tag{4.2.73}$$

The exact angular frequency for this nonlinear oscillator is $A^{1/2}\Omega_{\text{exact}} = 1.110$, while our calculation gives $A^{1/2}\Omega = 1.075$. Therefore, the percentage error is 3.1%.

4.3 Discussion

We complete this chapter by commenting briefly on some of the advantages and difficulties associated with the parameter expansion method for calculating analytic approximations to the periodic solutions of oscillatory systems. The basis of this method rests on expanding "numbers" or terms in the equation in a power (asymptotic) series in a parameter p and then carrying out a perturbation calculation under the assumption that p is small, i.e., $0 < p \ll 1$. Next, the calculated expressions are evaluated at $p = 1$ and the further assumption is made that these results provide a solution to the original problem.

4.3.1 *Advantages*

- Parameter expansions may be applied to both standard and TNL oscillator differential equations.
- The technique can also be used to analyze general linear and nonlinear ordinary and partial differential equation [15].
- Once the parameter expansion modification is made to the equation, the calculation of the periodic solution and the period is straightforward and proceeds in exactly the same manner as standard perturbation methods.

4.3.2 *Difficulties*

- It takes skill (and some luck) to formulate the appropriate parameter expansion construction for a given oscillator equation. In particular, more than one p-expansion rewriting of the original differential equation exists, and except by explicit calculation of each formulation and comparing the results, no *a priori* principle currently exists to aid with this process.
- It is not clear how to proceed with the application of p-expansion methods for terms of order p^2 and higher. No calculation to date provides information as to how this can be accomplished without ambiguities in either the formulation and/or the procedures to calculate the angular frequencies and solutions.
- The p-expansion methodology does not permit its application to oscillatory systems having solutions with transient behavior, i.e., the amplitudes and frequencies depend on time. One consequence of this limitation is that a full investigation of systems having limit-cycles cannot take place.

Problems

4.1 What mathematical and/or physical interpretation can be given to the assumed expansions of the expressions in Eqs. (4.2.4) and (4.2.5)?

4.2 Is it possible to construct a consistent order p^2 expansion? If so, do this for the TNL oscillator

$$\ddot{x} + x^3 = 0.$$

Can this procedure be generalized to order p^k, where $k > 2$?

4.3 What are the advantages and difficulties that emerge from having more than one possible starting form for the differential equation being solved by application of the parameter expansion method?

4.4 Can the parameter expansion method be applied to

$$\ddot{x} + x^{1/3} = \epsilon(1 - x^2)\dot{x}, \quad 0 < \epsilon \ll 1?$$

4.5 Complete the details to obtain the final results in Eq. (4.2.64).

4.6 Construct an order p solution for the following TNL modification of the simple harmonic oscillator equation

$$\ddot{x} + x + x^{1/3} = 0.$$

4.7 The quadratic, nonlinear oscillator is

$$\ddot{x} + |x|x = 0$$

or

$$\ddot{x} + x^2 \text{sgn}(x) = 0.$$

What are the solutions for this equation based on the parameter expansion method? In particular, compare the calculated and exact values of the angular frequencies.

References

[1] M. Senator and C. N. Bapat, *Journal of Sound and Vibration* **164**, 1 (1993).
[2] R. E. Mickens, *Journal of Sound and Vibration* **224**, 167 (1999).
[3] J. H. He, *International Journal of Nonlinear Mechanics* **37**, 309 (2002).
[4] J. H. He, *International Journal of Modern Physics* **20B**, 1141 (2006).
[5] D. H. Shou and J. H. He, *International Journal of Nonlinear Sciences and Numerical Simulation* **8**, 121 (2007).
[6] R. E. Mickens, *Nonlinear Oscillations* (Cambridge University Press, New York, 1991).
[7] R. E. Mickens, *Journal of Sound and Vibration* **244**, 563 (2001).
[8] T. Özis and A. Yildirim, *Computers and Mathematics with Applications* **54**, 1184 (2007).
[9] R. E. Mickens, *Journal of Sound and Vibration* **246**, 375 (2001).
[10] R. E. Mickens, *Journal of Sound and Vibration* **255**, 789 (2002).
[11] R. E. Mickens, *Journal of Sound and Vibration* **292**, 964 (2006).
[12] T. Özis and A. Yildirim, *Journal of Sound and Vibration* **306**, 372 (2007).
[13] R. E. Mickens, *Oscillations in Planar Dynamic Systems* (World Scientific, Singapore, 1996).
[14] T. Özis and A. Yildirim, *International Journal of Nonlinear Sciences and Numerical Integration* **8**, 243 (2007).
[15] The review paper of He [4] contains references to a number of these applications. In particular, see his references [44, 45, 46, 48, 62, 72].

Chapter 5

Iteration Methods

This chapter introduces the iteration method as a technique for calculating approximations to the periodic solutions of TNL oscillator differential equations. Section 5.1 discusses the general procedures required to formulate an iteration scheme. We derive two representations and denote them as direct and extended iteration methods. Sections 5.2 and 5.3, respectively, illustrate the application of these schemes to the same set of TNL equations. Finally, in Section 5.4, we provide a brief summary of the advantages and disadvantages of iteration methods.

While our presentation is only concerned with TNL oscillator systems, the general methodology of iteration procedures can also be applied to standard nonlinear oscillator differential equations having the form

$$\ddot{x} + x = \epsilon f(x, \dot{x}),$$

where ϵ is a parameter.

5.1 General Methodology

The 1987 paper by Mickens provided a general basis for iteration methods as they are currently used in the calculation of approximations to the periodic solutions of nonlinear oscillatory differential equations. A generalization of this work was then given by Lim and Wu [2] and this was followed by an additional extension in Mickens [3].

5.1.1 *Direct Iteration*

The basic foundation behind iteration methods is to re-express the original nonlinear differential equation into one that involves solving an infinite sequence of linear equations, each of which can be solved, and such that at a particular stage of the calculation, knowledge of the solutions of the previous members of the sequence is required to solve the differential equation at that stage. The major issue is how to reformulate the original nonlinear differential such that a viable iteration scheme can be constructed. The following is an outline of what must be achieved in order to attain this goal:

1) Assume that the differential equation of interest is
$$F(\ddot{x}, x) = 0, \quad x(0) = A, \quad \dot{x}(0) = 0, \tag{5.1.1}$$
and further assume that it can be rewritten to the form
$$\ddot{x} + f(\ddot{x}, x) = 0. \tag{5.1.2}$$
2) Next, add $\Omega^2 x$ to both sides to obtain
$$\ddot{x} + \Omega^2 x = \Omega^2 x - f(x, \ddot{x}) \equiv G(x, \ddot{x}), \tag{5.1.3}$$
where the constant Ω^2 is currently unknown.
3) Now, formulate the iteration scheme in the following way
$$\ddot{x}_{k+1} + \Omega_k^2 x_{k+1} = G(x_k, \ddot{x}_k); \quad k = 0, 1, 2, \ldots, \tag{5.1.4}$$
with
$$x_0(t) = A\cos(\Omega_0 t), \tag{5.1.5}$$
such that the x_{k+1} satisfy the initial conditions
$$x_{k+1}(0) = A, \quad \dot{x}_{k+1}(0) = 0. \tag{5.1.6}$$
4) At each stage of the iteration, Ω_k is determined by the requirement that secular terms [4, 5] should not occur in the full solution of $x_{k+1}(t)$.
5) This procedure gives a sequence of solutions: $x_0(t), x_1(t), \ldots$. Since all solutions are obtained from solving linear equations, they are, in principle, easy to calculate. The only difficulty might be the algebraic intensity required to complete the calculations.

At this point, the following observations should be noted:

(i) The solution for $x_{k+1}(t)$ depends on having the solutions for k less than $(k+1)$.
(ii) The linear differential equation for $x_{k+1}(t)$ allows the determination of Ω_k by the requirement that secular terms be absent. Therefore, the angular frequency, "Ω," appearing on the right-hand side of Eq. (5.1.4) in the function $x_k(t)$, is Ω_k.
(iii) In general, if Eq. (5.1.2) is of odd parity, i.e.,
$$f(-\ddot{x}, -x) = -f(\ddot{x}, x),$$
then the $x_k(t)$ will only contain odd multiples of the angular frequency [6].

5.1.2 *Extended Iteration*

Consider the following generalization of Eq. (5.1.2),

$$\ddot{x} + f(\ddot{x}, \dot{x}, x) = 0, \quad x(0) = A, \quad \dot{x}(0) = 0, \tag{5.1.7}$$

where evidence exists for periodic solutions. Rewrite this equation to the form

$$\ddot{x} + \Omega^2 x = \Omega^2 x - f(\ddot{x}, \dot{x}, x) \equiv G(\ddot{x}, \dot{x}, x). \tag{5.1.8}$$

The proposed "extended iteration" scheme is

$$\begin{aligned}\ddot{x}_{k+1} + \Omega_k^2 x_{k+1} &= G(\ddot{x}_{k-1}, \dot{x}_{k-1}, x_{k-1}) \\ &\quad + G_x(\ddot{x}_{k-1}, \dot{x}_{k-1}, x_{k-1})(x_k - x_{k-1}) \\ &\quad + G_{\dot{x}}(\ddot{x}_{k-1}, \dot{x}_{k-1}, x_{k-1})(\dot{x}_k - \dot{x}_{k-1}) \\ &\quad + G_{\ddot{x}}(\ddot{x}_{k-1}, \dot{x}_{k-1}, x_{k-1})(\ddot{x}_k - \ddot{x}_{k-1})\end{aligned} \tag{5.1.9}$$

where

$$G_x = \frac{\partial G}{\partial x}, \quad G_{\dot{x}} = \frac{\partial G}{\partial \dot{x}}, \quad G_{\ddot{x}} = \frac{\partial G}{\partial \ddot{x}}, \tag{5.1.10}$$

and $x_{k+1}(t)$ must satisfy the initial conditions

$$x_{k+1}(0) = A, \quad \dot{x}_{k+1}(0) = 0. \tag{5.1.11}$$

Examination of Eq. (5.1.9) shows that it requires a knowledge of two "starter solutions," $x_{-1}(t)$ and $x_0(t)$. These are taken to be [2, 3]

$$x_{-1}(t) = x_0(t) = A\cos(\Omega_0 t). \tag{5.1.12}$$

The right-hand side of Eq. (5.1.9) is essentially the first term in a Taylor series expansion of the function $G(\ddot{x}_k, \dot{x}_k, x_k)$ at the point $(\ddot{x}_{k-1}, \dot{x}_{k-1}, x_{k-1})$ [7]. To illustrate this point, note that

$$x_k = x_{k-1} + (x_k - x_{k-1}) \tag{5.1.13}$$

and for some function $H(x)$, we have

$$\begin{aligned}H(x_k) &= H[x_{k-1} + (x_k - x_{k-1})] \\ &= H(x_{k-1}) + H_x(x_{k-1})(x_k - x_{k-1}) + \cdots.\end{aligned} \tag{5.1.14}$$

An alternative, but very insightful, modification of the above scheme was proposed by Hu [8]. In place of Eq. (5.1.13) use

$$x_k = x_0 + (x_k - x_0). \tag{5.1.15}$$

Then, Eq. (5.1.14) is changed to

$$H(x_k) = H[x_0 + (x_k - x_0)] = H(x_0) + H_x(x_0)(x_k - x_0) + \cdots \tag{5.1.16}$$

and the corresponding modification to Eq. (5.1.9) is

$$\begin{aligned}\ddot{x}_{k+1} + \Omega_k^2 x_{k+1} &= G(\ddot{x}_0, \dot{x}_0, x_0) + G_x(\ddot{x}_0, \dot{x}_0, x_0)(x_k - x_0)\\ &\quad + G_{\dot{x}}(\ddot{x}_0, \dot{x}_0, x_0)(\dot{x}_k - \dot{x}_0) + G_{\ddot{x}}(\ddot{x}_0, \dot{x}_0, x_0)(\ddot{x}_k - \ddot{x}_0). \qquad (5.1.17)\end{aligned}$$

The latter scheme is computationally easier to work with, for $k \geq 2$, than the one given in Eq. (5.1.9). The essential idea is that if $x_0(t)$ is a good approximation, then the expansion should take place at $x = x_0$. Also, as pointed out by Hu [8], the x_0 in $(x_k - x_0)$ is not the same for all k. In particular, x_0 in $(x_1 - x_0)$ is the function $A\cos(\Omega_1 t)$, while the x_0 in $(x_2 - x_0)$ is $A\cos(\Omega_2 t)$.

The next two sections apply both of these iteration schemes to the same set of TNL oscillator differential equations. These applications will allow the subtleties of iteration methods to be understood.

5.2 Worked Examples: Direct Iteration

In all of the calculations to follow, the initial conditions for the solutions of the appropriate differential equations are taken to be

$$x(0) = A, \quad \dot{x}(0) = 0. \qquad (5.2.1)$$

Similarly, $x_0(t)$ is

$$x_0(t) = A\cos(\Omega_0 t). \qquad (5.2.2)$$

5.2.1 $\dot{x} + x^3 = 0$

A possible iteration scheme for this equation is

$$\ddot{x}_{k+1} + \Omega_k^2 x_{k+1} = \Omega_k^2 x_k - x_k^3. \qquad (5.2.3)$$

For $k = 0$, we have

$$\begin{aligned}\ddot{x}_1 + \Omega_0^2 x_1 &= \Omega_0^2 x_0 - x_0^3 = \Omega_0^2(A\cos\theta) - (A\cos\theta)^3\\ &= \left[\Omega_0^2 - \left(\frac{3}{4}\right)A^2\right]A\cos\theta - \left(\frac{A^3}{4}\right)\cos 3\theta, \qquad (5.2.4)\end{aligned}$$

where $\theta = \Omega_0 t$. To derive this result use was made of the following trigonometric relation

$$(\cos\theta)^3 = \left(\frac{3}{4}\right)\cos\theta + \left(\frac{1}{4}\right)\cos 3\theta.$$

Secular terms will not appear in the solution for $x_1(t)$ if the coefficient of the $\cos\theta$ term is zero, i.e.,

$$\Omega_0^2 - \left(\frac{3}{4}\right) A^3 = 0, \tag{5.2.5}$$

and

$$\Omega_0(A) = \left(\frac{3}{4}\right)^{1/2} A. \tag{5.2.6}$$

Under the no secular term requirement, Eq. (5.2.4) reduces to

$$\ddot{x}_1 + \Omega_0^2 x_1 = -\left(\frac{A^3}{4}\right)\cos 3\theta. \tag{5.2.7}$$

The particular solution for this equation takes the form

$$x_1^{(p)}(t) = D\cos(3\theta).$$

Substitution of this into Eq. (5.2.7) gives

$$(-9\Omega_0^2 + \Omega_0^2)D = -\left(\frac{A^3}{4}\right)$$

and

$$D = \frac{A^3}{32\Omega_0^2} = \left(\frac{A^3}{32}\right)\left(\frac{4}{3A^2}\right) = \frac{A}{24}.$$

Therefore, the full solution to Eq. (5.2.7) is

$$x_1(t) = x_1^{(h)} + x_1^{(p)} = C\cos\theta + \left(\frac{A}{24}\right)\cos 3\theta,$$

where $C\cos\theta$ is the solution to the homogeneous equation

$$\ddot{x}_1 + \Omega_0^2 x_1 = 0. \tag{5.2.8}$$

Since $x_1(0) = A$, then

$$A = C + \left(\frac{A}{24}\right)$$

or

$$C = \left(\frac{23}{24}\right) A,$$

and the full solution to Eq. (5.2.7) is

$$x_1(t) = A\left[\left(\frac{23}{24}\right)\cos\theta + \left(\frac{1}{24}\right)\cos 3\theta\right]. \tag{5.2.9}$$

If we stop the calculation at this point, then the first-approximation to the periodic solution is

$$x_1(t) = A\left[\left(\frac{23}{24}\right)\cos\left(\sqrt{\frac{3}{4}}\,At\right) + \left(\frac{1}{24}\right)\cos\left(3\sqrt{\frac{3}{4}}\,At\right)\right]. \tag{5.2.10}$$

However, to extend our calculation to the next level, $x_1(t)$ takes the form given by Eq. (5.2.9), but θ is now equal to $\Omega_1 t$, i.e.,

$$\begin{aligned} x_1(t) &= A\left[\left(\frac{23}{24}\right)\cos(\Omega_1 t) + \left(\frac{1}{24}\right)\cos(3\Omega_1 t)\right] \\ &= A\left[\left(\frac{23}{24}\right)\cos\theta + \left(\frac{1}{24}\right)\cos 3\theta\right]. \end{aligned} \tag{5.2.11}$$

Note, we denote the phase of the trigonometric expressions by θ, i.e., $\theta = \Omega_1 t$. This short-hand notation will be used for the remainder of the chapter.

The next approximation, $x_2(t)$, requires the solution to

$$\ddot{x}_2 + \Omega_1^2 x_2 = \Omega_1^2 x_1 - x_1^3. \tag{5.2.12}$$

We now present the full details on how to evaluate the right-hand side of Eq. (5.2.12). These steps demonstrate what must be done for this type of calculation. In the calculations for other TNL oscillators, we will generally omit many of the explicit details contained in this section.

To begin, consider the following result

$$\begin{aligned} (a_1\cos\theta + a_2\cos 3\theta)^3 &= (a_1\cos\theta)^3 + 3(a_1\cos\theta)^2(a_2\cos 3\theta) \\ &\quad + 3(a_1\cos\theta)(a_2\cos 3\theta)^2 + (a_2\cos 3\theta)^3. \end{aligned}$$

Using

$$(\cos\theta_1)(\cos\theta_2) = \left(\frac{1}{2}\right)[\cos(\theta_1+\theta_2) + \cos(\theta_1-\theta_2)]$$

and the previous expression for $(\cos\theta_1)^3$, we find

$$\begin{aligned} (a_1\cos\theta + a_2\cos 3\theta)^3 &= f_1\cos\theta + f_2\cos 3\theta \\ &\quad + f_3\cos 5\theta + f_4\cos 7\theta + f_5\cos 9\theta \end{aligned} \tag{5.2.13}$$

where

$$\begin{cases} f_1 = \left(\dfrac{3}{4}\right)[a_1^3 + a_1^2 a_2 + 2a_1 a_2^2], \\ f_2 = \left(\dfrac{1}{4}\right)[a_1^3 + 6a_1^2 a_2 + 3a_2^3], \\ f_3 = \left(\dfrac{3}{4}\right)[a_1^2 a_2 + a_1 a_2^2], \\ f_4 = \left(\dfrac{3}{4}\right)a_1 a_2^2, \\ f_5 = \dfrac{a_2^3}{4}. \end{cases} \tag{5.2.14}$$

For our problem, see Eq. (5.2.11), we have

$$\begin{cases} a_1 = \left(\dfrac{23}{24}\right) A \equiv \alpha A, \\ a_2 = \left(\dfrac{1}{24}\right) A \equiv \beta A. \end{cases} \tag{5.2.15}$$

Using these results, Eq. (5.2.12) becomes

$$\begin{aligned} \ddot{x}_2 + \Omega_1^2 x_2 = (\Omega_1^2 a_1 - f_1)\cos\theta + (\Omega_1^2 a_2 - f_2)\cos 3\theta \\ - f_3 \cos 5\theta - f_4 \cos 7\theta - f_5 \cos 9\theta. \end{aligned} \tag{5.2.16}$$

Secular terms may be eliminated in the solution for $x_2(t)$ if the coefficient of the $\cos\theta$ term is zero, i.e.,

$$\Omega_1^2 a_1 - f_1 = 0, \tag{5.2.17}$$

and

$$\begin{aligned} \Omega_1^2(A) &= \frac{f_1}{a_1} = \left(\frac{3}{4}\right)[\alpha^3 + \alpha^2\beta + 2\alpha\beta^2]A^3/\alpha A \\ &= \left[\left(\frac{3}{4}\right)A^3\right][\alpha^2 + \alpha\beta + 2\beta^2] = \Omega_0^2(A)h(\alpha,\beta), \end{aligned} \tag{5.2.18}$$

where

$$h(\alpha,\beta) = \alpha^2 + \alpha\beta + 2\beta^2. \tag{5.2.19}$$

Examination of Eqs. (5.2.18) and (5.2.19) shows that $h(\alpha,\beta)$ provides a correction to the square of the first-order angular frequency, $\Omega_0^2(A)$. Since $\alpha = 23/24$ and $\beta = 1/24$, then

$$\Omega_0(A) = \sqrt{\frac{3}{4}}\,A = (0.866025)A, \tag{5.2.20}$$

$$\Omega_1(A) = (0.849326)A. \tag{5.2.21}$$

These are to be compared to

$$\Omega_{\text{exact}}(A) = (0.847213)A. \tag{5.2.22}$$

The corresponding percentage errors are

$$\left|\frac{\Omega_{\text{exact}} - \Omega_0}{\Omega_{\text{exact}}}\right| \cdot 100 = 2.2\%, \quad \left|\frac{\Omega_{\text{exact}} - \Omega_1}{\Omega_{\text{exact}}}\right| \cdot 100 = 0.2\%. \tag{5.2.23}$$

Let us now calculate $x_2(t)$. This function is a solution to

$$\begin{aligned} \ddot{x}_2 + \Omega_1^2 x_2 = (\Omega_1^2 a_2 - f_2)\cos 3\theta - f_3 \cos 5\theta \\ - f_4 \cos 7\theta - f_5 \cos 9\theta. \end{aligned} \tag{5.2.24}$$

The particular solution is

$$x_2^{(p)}(t) = D_1 \cos 3\theta + D_2 \cos 5\theta + D_3 \cos 7\theta + D_4 \cos 9\theta \qquad (5.2.25)$$

where (D_1, D_2, D_3, D_4) are constants that can be found by substituting $x_2^{(p)}$ into Eq. (5.2.24) and equating similar terms on both the left and right sides. Performing this procedure gives

$$\begin{aligned}
D_1 &= \frac{\Omega_1^2 a_2 - f_2}{(-8)\Omega_1^2} \\
&= -\left(\frac{A}{24}\right)\left[\frac{3\beta(\alpha^2+\alpha\beta+2\beta^2)-(\alpha^3+6\alpha^2\beta+3\beta^3)}{\alpha^2+\alpha\beta+2\beta^2}\right], \\
D_2 &= \frac{f_3}{24\Omega_1^2} = \left(\frac{A}{24}\right)\left[\frac{\alpha^2\beta+\alpha\beta^2}{\alpha^2+\alpha\beta+2\beta^2}\right], \\
D_3 &= \frac{f_4}{48\Omega_1^2} = \left(\frac{A}{48}\right)\left[\frac{\alpha\beta^2}{\alpha^2+\alpha\beta+2\beta^2}\right], \\
D_4 &= \frac{f_5}{80\Omega_1^2} = \left(\frac{A}{240}\right)\left[\frac{\beta^3}{\alpha^2+\alpha\beta+2\beta^2}\right].
\end{aligned}$$

In these expressions, we have replaced Ω_1^2 by the results in Eqs. (5.2.18) and (5.2.19).

The complete solution for $x_2(t)$ is

$$x_2(t) = x_2^{(H)}(t) + x_2^{(p)} = C\cos\theta + x_2^{(p)}.$$

For $t = 0$, we have

$$A = C + (D_1 + D_2 + D_3 + D_4).$$

If we define

$$D_i = A\bar{D}_i; \quad i = 1, 2, 3, 4;$$

then

$$C = 1 - (\bar{D}_1 + \bar{D}_2 + \bar{D}_3 + \bar{D}_4)A,$$

and

$$\begin{aligned}
x_2(t) = {} & [1 - (\bar{D}_1 + \bar{D}_2 + \bar{D}_3 + \bar{D}_4)]A\cos\theta \\
& + A[\bar{D}_1 \cos 3\theta + \bar{D}_2 \cos 5\theta + \bar{D}_3 \cos 7\theta + \bar{D}_4 \cos 9\theta],
\end{aligned}$$

where $\theta = \Omega_1(A)t$.

Using the numerical values for α and β, the $\bar{D}$'s can be calculated; we find their values to be

$$\bar{D}_1 = 0.042876301 \approx (4.29) \cdot 10^{-2},$$

$$\bar{D}_2 = 0.001729754 \approx (1.73) \cdot 10^{-3},$$
$$\bar{D}_3 = 0.000036038 \approx (3.60) \cdot 10^{-5},$$
$$\bar{D}_4 = 0.000000313 \approx (3.13) \cdot 10^{-7}.$$

Therefore, we have for $x_2(t)$ the expression

$$\begin{aligned} x_2(t) &= A\big[(0.955)\cos\theta + (4.29)\cdot 10^{-2}\cos 3\theta \\ &\quad + (1.73)\cdot 10^{-3}\cos 5\theta + (3.60)\cdot 10^{-5}\cos 7\theta \\ &\quad + (3.13)\cdot 10^{-7}\cos 9\theta\big], \\ \theta &= \Omega_1(A)t = (0.849325713)A. \end{aligned} \tag{5.2.26}$$

Note that the ratios of neighboring coefficients are

$$\frac{a_1}{a_0} \approx (4.49)\cdot 10^{-2}, \qquad \frac{a_2}{a_1} \approx (4.03)\cdot 10^{-2}$$
$$\frac{a_3}{a_2} \approx (2.08)\cdot 10^{-2}, \qquad \frac{a_4}{a_3} \approx (0.87)\cdot 10^{-2}.$$

These results indicate that the coefficients decrease rapidly, dropping by approximately two orders of magnitude from one coefficient to its next neighbor [9]; see Mickens [9, Section 4.2.1].

5.2.2 $\ddot{x} + x^3/(1 + x^2) = 0$

This equation

$$\ddot{x} + \frac{x^3}{1 + x^2} = 0, \tag{5.2.27}$$

can be rewritten as follows

$$(1 + x^2)\ddot{x} + x^3 = 0$$

$$\ddot{x} = -x^3 - x^2\ddot{x}$$

$$\ddot{x} + \Omega^2 x = \Omega^2 x - x^3 - x^2\ddot{x}.$$

Therefore, an associated iteration scheme is

$$\ddot{x}_{k+1} + \Omega_k^2 x_{k+1} = \Omega_k^2 x_k - x_k^3 - x_k^2\ddot{x}_k. \tag{5.2.28}$$

For $k = 0$, we have

$$\ddot{x}_1 + \Omega_0^2 x_1 = \Omega_0^2 x_0 - x_0^3 - x_0^2\ddot{x}_0, \tag{5.2.29}$$

where $x_0(t) = A\cos(\Omega_2 t) = A\cos\theta$. With $x_0(t)$ substituted into the right-hand side, Eq. (5.2.29) becomes

$$\ddot{x}_1 + \Omega_0^2 x_1 = \Omega_0^2(A\cos\theta) - (A\cos\theta)^3 - (A\cos\theta)^2(-\Omega_0^2 A\cos\theta)$$

$$= \left[\Omega_0^2 - \frac{3A^2}{4} + \frac{3A^2\Omega_0^2}{4}\right] A\cos\theta + \left[\frac{\Omega_0^2 A^3}{4} - \frac{A^3}{4}\right]\cos 3\theta. \tag{5.2.30}$$

Secular terms can be eliminated if the coefficient of the $\cos\theta$ term is set to zero, i.e.,

$$\Omega_0^2 - \frac{3A^2}{4} + \frac{3A^2\Omega_0^2}{4} = 0,$$

or

$$\Omega_0^2(A) = \frac{\left(\frac{3A^2}{4}\right)}{1 + \left(\frac{3A^2}{4}\right)}. \tag{5.2.31}$$

This result can be used to evaluate the coefficient of the $\cos 3\theta$ term, i.e.,

$$\frac{\Omega_0^2 A^3}{4} - \frac{A^3}{4} = \left(\frac{A^3}{4}\right)(\Omega_0^2 - 1) = -\left(\frac{\Omega_0^2}{3}\right)A.$$

With this evaluation of the coefficient, the differential equation for $x_1(t)$ is

$$\ddot{x}_1 + \Omega_0^2 x_1 = -\left(\frac{\Omega_0^2}{3}\right)A\cos 3\theta. \tag{5.2.32}$$

The particular solution is $x_1^{(p)}(t) = D\cos 3\theta$, where D is determined by substituting $x_1^{(p)}(t)$ into Eq. (5.2.32), i.e.,

$$(-9\Omega_0^2 + \Omega_0^2)D = -\left(\frac{\Omega_0^2}{3}\right)A$$

or

$$D = \frac{A}{24}. \tag{5.2.33}$$

Since

$$x_1(t) = x_1^{(H)}(t) + x_1^{(p)} = C\cos\theta + \left(\frac{A}{24}\right)\cos 3\theta,$$

the $x_1(0) = A$ gives

$$C + \frac{A}{24} = A \quad \text{or} \quad C = \left(\frac{23}{24}\right)A,$$

and

$$x_1(t) = A\left[\left(\frac{23}{24}\right)\cos\theta + \left(\frac{1}{24}\right)\cos 3\theta\right]. \tag{5.2.34}$$

If we terminate the calculation at this point, then

$$\begin{cases} x_1(t) = A\left[\left(\frac{23}{24}\right)\cos(\Omega_0 t) + \left(\frac{1}{24}\right)\cos(3\Omega_0 t)\right], \\ \Omega_0^2 = \frac{\left(\frac{3A^2}{4}\right)}{1+\left(\frac{3A^2}{4}\right)}. \end{cases} \tag{5.2.35}$$

However, continuing to the next level of the iteration scheme gives

$$\ddot{x}_2 + \Omega_1^2 x_2 = \Omega_1^2 x_1 - x_1^3 - x_1^2 \ddot{x}_1, \tag{5.2.36}$$

where $x_1(t)$, on the right-hand side of this equation, is

$$x_1(t) = A\left[\left(\frac{23}{24}\right)\cos\theta + \left(\frac{1}{24}\right)\cos 3\theta\right], \quad \theta = \Omega_1 t. \tag{5.2.37}$$

(Note, θ depends on Ω_1 and not Ω_0.) If this $x_1(t)$ is substituted into Eq. (5.2.36), then after some trigonometric and algebraic manipulations, the following result is found

$$\begin{aligned} \ddot{x}_2 + \Omega_1^2 x_2 = {} & \left\{\Omega_1^2 - \left(\frac{3A^2}{4}\right)(\alpha^2 + \alpha\beta + 2\beta^2) \right. \\ & \left. + \Omega_1^2\left(\frac{3A^2}{4}\right)\left[\alpha^2 + \left(\frac{11}{3}\right)\alpha\beta + \left(\frac{38}{3}\right)\beta^2\right]\right\} A\alpha\cos\theta \\ & + \text{HOH}, \end{aligned} \tag{5.2.38}$$

where HOH = higher-order harmonics and

$$\alpha = \frac{23}{24}, \quad \beta = \frac{1}{24}. \tag{5.2.39}$$

Therefore,

$$\begin{aligned} h_1(\alpha,\beta) &= \alpha^2 + \alpha\beta + 2\beta^2 = 0.961806, \\ h_2(\alpha,\beta) &= \alpha^2 + \left(\frac{11}{3}\right)\alpha\beta + \left(\frac{38}{3}\right)\beta^2 = 1.086805, \end{aligned}$$

and Ω_1^2 can be determined by requiring the coefficient of $\cos\theta$, in Eq. (5.2.38), to be zero. Carrying out this task gives

$$\Omega_1^2(A) = \frac{\left(\frac{3A^2}{4}\right)h_1}{1 + \left(\frac{3A^2}{4}\right)h_2} = \frac{(0.9618)\left(\frac{3A^2}{4}\right)}{1 + (1.0868)\left(\frac{3A^2}{4}\right)}. \tag{5.2.40}$$

Comparing $\Omega_0^2(A)$, from Eq. (5.2.31), with the above evaluation for $\Omega_1^2(A)$, we find that $\Omega_1^2(A)$ is a minor modification of the form given for $\Omega_0^2(A)$.

Finally, it should be observed that the Duffing-harmonic oscillator, Eq. (5.2.27), has the following properties

$$x \text{ small} : \ddot{x} + x^3 \simeq 0, \quad \Omega_{\text{exact}}(A) = (0.8472)A,$$
$$x \text{ large} : \ddot{x} + x \simeq 0, \quad \Omega_{\text{exact}}(A) = 1,$$

while from Eqs. (5.2.31) and (5.2.40), we have

$$x \text{ small} : \Omega_0(A) = (0.8660)A, \quad \Omega_1(A) = (0.8493)A,$$
$$x \text{ large} : \Omega_0(A) = 1, \qquad \Omega_1(A) = 0.9407.$$

Therefore, $\Omega_0(A)$ gives the correct value of $\Omega(A)$ for large A, while $\Omega_1(A)$ gives the better estimate for small A.

5.2.3 $\ddot{x} + x^{-1} = 0$

This TNL oscillator differential equation can be written as

$$x\ddot{x} + 1 = 0$$

$$\ddot{x} = -(\ddot{x})^2 x$$

$$\ddot{x} + \Omega^2 x = \Omega^2 x - (\ddot{x})^2 x.$$

This last expression suggests the following iteration scheme

$$\ddot{x}_{k+1} + \Omega_k^2 x_{k+1} = \Omega_k^2 x_k - (\ddot{x}_k)^2 x_k. \tag{5.2.41}$$

For $k = 0$ and $x_0(t) = A\cos\theta$, $\theta = \Omega_0 t$, we have

$$\begin{aligned}\ddot{x}_1 + \Omega_0^2 x_1 &= (\Omega_0^2 A\cos\theta) - (-\Omega_0^2 A\cos\theta)^2 (A\cos\theta)\\ &= \Omega_0^2\left[1 - \frac{3A^2\Omega_0^2}{4}\right] A\cos\theta - \left(\frac{A^3\Omega_0^4}{4}\right)\cos 3\theta.\end{aligned} \tag{5.2.42}$$

The elimination of secular terms gives

$$1 - \frac{3A^2\Omega_0^2}{4} = 0,$$

and

$$\Omega_0^2(A) = \left(\frac{4}{3}\right)\frac{1}{A^2}. \tag{5.2.43}$$

Therefore, $x_1(t)$ satisfies the equation

$$\ddot{x}_1 + \Omega_0^2 x_1 = -\left(\frac{A^3\Omega_0^4}{4}\right)\cos 3\theta. \tag{5.2.44}$$

The particular solution, $x_1^{(p)}(t)$, is

$$x_1^{(p)}(t) = \left(\frac{A^3\Omega_0^2}{32}\right)\cos 3\theta = \left(\frac{A}{24}\right)\cos 3\theta.$$

Therefore, the full solution is

$$x_1(t) = C\cos\theta + \left(\frac{A}{24}\right)\cos 3\theta.$$

Using $x_1(0) = A$, then $C = 23/24$ and

$$x_1(t) = A\left[\left(\frac{23}{24}\right)\cos\theta + \left(\frac{1}{24}\right)\cos 3\theta\right]. \tag{5.2.45}$$

If the calculation is stopped at this point, then

$$\begin{cases} x_1(t) = \left[\left(\frac{23}{24}\right)\cos(\Omega_0 t) + \left(\frac{1}{24}\right)\cos(3\Omega_0 t)\right] \\ \Omega_0(A) = \frac{2}{\sqrt{3}A} = \frac{1.1547}{A}. \end{cases} \tag{5.2.46}$$

Note that [10]

$$\Omega_{\text{exact}}(A) = \frac{\sqrt{2\pi}}{2A} = \frac{1.2533141}{A}, \tag{5.2.47}$$

and

$$\left|\frac{\Omega_{\text{exact}} - \Omega_0}{\Omega_{\text{exact}}}\right| \cdot 100 = 7.9\% \text{ error}. \tag{5.2.48}$$

Proceeding to the second level of iteration, $x_2(t)$ must satisfy the equation

$$\ddot{x}_2 + \Omega_1^2 x_2 = \Omega_1^2 x_1 - (\ddot{x}_1)^2 x_1, \tag{5.2.49}$$

where

$$x_1(t) = A\left[\left(\frac{23}{24}\right)\cos(\Omega_1 t) + \left(\frac{1}{24}\right)\cos(3\Omega_1 t)\right]. \tag{5.2.50}$$

Let $\theta = \Omega_1 t$ and substitute this $x_1(t)$ into the right-hand side of Eq. (5.2.49); doing so gives

$$\ddot{x}_2 + \Omega_1^2 x_2 = \Omega_1^2\left[\alpha - \left(\frac{3}{4}\right)A^2\Omega_1^2 g(\alpha,\beta)\right]A\cos\theta + \text{HOH}, \tag{5.2.51}$$

where

$$g(\alpha,\beta) = \alpha^3 + \left(\frac{19}{3}\right)\alpha^2\beta + 66\alpha\beta^2 + 27\beta^3, \tag{5.2.52}$$

and

$$\alpha = \frac{23}{24}, \quad \beta = \frac{1}{24}. \tag{5.2.53}$$

The absence of secular terms gives

$$\Omega_1^2 = \left[\left(\frac{4}{3}\right)\frac{1}{A^2}\right]\left[\frac{\alpha}{g(\alpha,\beta)}\right], \tag{5.2.54}$$

and

$$\Omega_1(A) = \frac{1.0175}{A},$$

with

$$\left|\frac{\Omega_{\text{exact}} - \Omega_1}{\Omega_1}\right| \cdot 100 = 18.1\% \text{ error}. \tag{5.2.55}$$

The existence of such a large percentage-error suggests that we should try an alternative iteration scheme and determine if a better result can be found. This second scheme is

$$\ddot{x}_{k+1} + \Omega_k^2 x_{k+1} = \ddot{x}_k - \Omega_k^2 (x_k)^2 \ddot{x}_k. \tag{5.2.56}$$

For $k = 0$, we have

$$\ddot{x}_1 + \Omega_0^2 x_1 = \ddot{x}_0 - \Omega_0^2 (x_0)^2 \ddot{x}_0. \tag{5.2.57}$$

With $x_0(t) = A\cos(\Omega_0 t)$, we find that

$$\Omega_0(A) = \sqrt{\frac{4}{3}}\left(\frac{1}{A}\right),$$

which is exactly the same result as previously given in Eq. (5.2.46). Similarly, we also determine that $x_1(t)$ is

$$x_1(t) = A\left[\left(\frac{25}{24}\right)\cos\theta - \left(\frac{1}{24}\right)\cos 3\theta\right], \tag{5.2.58}$$

a result which differs from the previous calculation, i.e., compare the coefficients in Eqs. (5.2.45) and (5.2.58). Further, the value of $\Omega_1(A)$, for the iteration scheme of Eq. (5.2.56), is

$$\Omega_1^2(A) = \left[\left(\frac{4}{3}\right)\frac{1}{A^2}\right]\left[\frac{\alpha}{h(\alpha,\beta)}\right], \tag{5.2.59}$$

where, for this case,

$$\begin{cases} h(\alpha,\beta) = \alpha^3 - \left(\frac{11}{3}\right)\alpha^2\beta + \left(\frac{38}{3}\right)\alpha\beta^2, \\ \alpha = \frac{25}{24}, \quad \beta = \frac{1}{24}, \end{cases} \tag{5.2.60}$$

and

$$\Omega_1(A) = \frac{1.0262}{A}, \tag{5.2.61}$$

with

$$\left| \frac{\Omega_{\text{exact}} - \Omega_1(A)}{\Omega_{\text{exact}}} \right| \cdot 100 = 18\% \text{ error.} \tag{5.2.62}$$

The general conclusion reached is that if the percentage error in the angular frequency is to be taken as a measure of the accuracy of this calculation, then the iteration method does not appear to work well for this particular TNL oscillator. In fact, since the error for $\Omega_0(A)$ is less than that of $\Omega_1(A)$, the two schemes may give (increasing in value) erroneous results for the angular frequency as k becomes larger.

5.2.4 $\ddot{x} + sgn(x) = 0$

This equation models the antisymmetric, constant force oscillator. If we write it as

$$\text{sgn}(x) = -\ddot{x}, \tag{5.2.63}$$

and square both sides, then

$$1 = (\ddot{x})^2$$

$$x = (\ddot{x})^2 x$$

$$\ddot{x} + \Omega^2 x = \ddot{x} + \Omega^2 (\ddot{x})^2 x,$$

and this form suggests the following iteration scheme

$$\ddot{x}_{k+1} + \Omega_k^2 x_{k+1} = \ddot{x}_k + \Omega_k^2 (\ddot{x}_k)^2 x_k. \tag{5.2.64}$$

For $k = 0$ and $x_0(t) = A\cos\theta$, $\theta = \Omega_0 t$, then

$$\begin{aligned} \ddot{x}_1 + \Omega_0^2 x_1 &= \ddot{x}_0 + \Omega_0^2 (\ddot{x}_0)^2 x_0 \\ &= (-\Omega_0^2 A\cos\theta) + \Omega_0^2 (-\Omega_0^2 A\cos\theta)^2 (A\cos\theta) \\ &= -\Omega_0^2 \left[1 - \frac{3\Omega_0^4 A^2}{4} \right] A\cos\theta + \left(\frac{\Omega_0^6 A^3}{4} \right) \cos 3\theta. \end{aligned} \tag{5.2.65}$$

The no secular term condition gives

$$1 - \frac{3\Omega_0^4 A^2}{4} = 0$$

or

$$\Omega_0^4 = \left(\frac{4}{3}\right)\frac{1}{A^2}, \tag{5.2.66}$$

and

$$\Omega_0(A) = \left(\frac{4}{3}\right)^{1/4}\frac{1}{A^{1/2}} = \frac{1.0745699}{A}. \tag{5.2.67}$$

Using the exact value

$$\Omega_{\text{exact}}(A) = \left(\frac{\pi}{2\sqrt{2}}\right)\frac{1}{A^{1/2}} = \frac{1.1107207}{A^{1/2}}, \tag{5.2.68}$$

the percentage error for $\Omega_0(A)$ is

$$\left|\frac{\Omega_{\text{exact}} - \Omega_0}{\Omega_{\text{exact}}}\right| \cdot 100 = 3.3\% \text{ error}. \tag{5.2.69}$$

The full solution of

$$\ddot{x}_1 + \Omega_0^2 x_1 = \left(\frac{\Omega_0^6 A^3}{4}\right)\cos 3\theta,$$

with $x_1(0) = A$, is

$$x_1(t) = A\left[\left(\frac{25}{24}\right)\cos\theta - \left(\frac{1}{24}\right)\cos 3\theta\right]. \tag{5.2.70}$$

If we stop the calculation at this stage, then

$$x_1(t) = A\left[\left(\frac{25}{24}\right)\cos(\Omega_0 t) - \left(\frac{1}{24}\right)\cos(3\Omega_0 t)\right], \tag{5.2.71}$$

where $\Omega_0(A)$ is given in Eq. (5.2.67).

For $k = 1$, we have

$$\ddot{x}_2 + \Omega_1^2 x_2 = \ddot{x}_1 + \Omega_1^2 x_1 (\ddot{x}_1)^2, \tag{5.2.72}$$

where

$$\begin{cases} x_1(t) = A\left[\left(\frac{25}{24}\right)\cos\theta - \left(\frac{1}{24}\right)\cos 3\theta\right], \\ \theta = \Omega_1 t. \end{cases} \tag{5.2.73}$$

Substituting Eq. (5.2.73) into Eq. (5.2.72) and simplifying gives the result

$$\ddot{x}_2 + \Omega_1^2 x_2 = -\Omega_1^2\left[\alpha - \left(\frac{3\Omega_1^4 A^2}{4}\right)h(\alpha,\beta)\right]A\cos\theta + \text{HOH}, \tag{5.2.74}$$

where

$$h(\alpha,\beta) = \alpha^3 - \left(\frac{19}{3}\right)\alpha^2\beta + \left(\frac{198}{3}\right)\alpha\beta^2, \tag{5.2.75}$$

with

$$\alpha = \frac{25}{24}, \quad \beta = \frac{1}{24}. \tag{5.2.76}$$

Setting the coefficient of the $\cos\theta$ term to zero, gives

$$\Omega_1^4 = \left[\left(\frac{4}{3}\right)\frac{1}{A^2}\right]\left[\frac{\alpha}{h(\alpha,\beta)}\right] \tag{5.2.77}$$

and, on evaluating the right-hand side

$$\Omega_1(A) = \frac{1.095788}{A^{1/2}}. \tag{5.2.78}$$

Therefore, $\Omega_1(A)$ has a percentage error of 1.3%, as compared to $\Omega_0(A)$ where the error is 3.3%.

5.2.5 $\ddot{x} + x^{1/3} = 0$

The cube-root TNL oscillator equation can be rewritten as

$$x = -(\ddot{x})^3$$

$$\ddot{x} + \Omega^2 x = \ddot{x} - \Omega^2(\ddot{x})^3$$

and the associated iteration scheme is

$$\ddot{x}_{k+1} + \Omega_k^2 x_{k+1} = \ddot{x}_k - \Omega_k^2(\ddot{x}_k)^2. \tag{5.2.79}$$

With $x_0(t) = A\cos(\Omega_0 t) = A\cos\theta$, we have

$$\ddot{x}_1 + \Omega_0^2 x_0 = -(\Omega_0^2)\left[1 - \Omega_0^6\left(\frac{3A^2}{4}\right)\right]A\cos\theta + \left(\frac{\Omega_0^8 A^3}{4}\right)\cos 3\theta. \tag{5.2.80}$$

The elimination of secular terms in the solution $x_1(t)$ gives

$$1 - \Omega_0^6\left(\frac{3A^2}{4}\right) = 0,$$

or

$$\Omega_0^6(A) = \left(\frac{4}{3}\right)\left(\frac{1}{A^2}\right),$$

and

$$\Omega_0(A) = \left(\frac{4}{3}\right)^{1/6}\frac{1}{A^{1/3}} = \frac{1.0491151}{A^{1/3}}. \tag{5.2.81}$$

Since the exact value for $\Omega(A)$ for the cube-root equation is

$$\Omega_{\text{exact}}(A) = \frac{1.070451}{A^{1/3}}, \tag{5.2.82}$$

then

$$\left|\frac{\Omega_{\text{exact}} - \Omega_0}{\Omega_{\text{exact}}}\right| \cdot 100 = 2.0\% \text{ error}. \tag{5.2.83}$$

The full solution to

$$\ddot{x}_1 + \Omega_0^2 x_1 = \left(\frac{\Omega_0^8 A^3}{4}\right) \cos 3\theta,$$

is

$$x_1(t) = A\left[\left(\frac{25}{24}\right)\cos\theta - \left(\frac{1}{24}\right)\cos 3\theta\right], \tag{5.2.84}$$

and if we stop at this level of calculation, it follows that $x_1(t)$ is

$$x_1(t) = A\left[\left(\frac{25}{24}\right)\cos(\Omega_0 t) - \left(\frac{1}{24}\right)\cos(3\Omega_0 t\right], \tag{5.2.85}$$

where $\Omega_0(A)$ is the expression given in Eq. (5.2.81).

At the next level of the calculation, i.e., for $k = 1$, we have

$$\ddot{x}_2 + \Omega_1^2 x_2 = \ddot{x}_1 - \Omega_1^2 (\ddot{x}_1)^3. \tag{5.2.86}$$

If

$$\begin{cases} x_1(t) = A[\alpha\cos\theta - \beta\cos 3\theta], \\ \theta = \Omega_1 t, \quad \alpha = \dfrac{25}{24}, \quad \beta = \dfrac{1}{24}, \end{cases} \tag{5.2.87}$$

is substituted into the right-hand side of Eq. (5.2.86), then the resulting expression is

$$\ddot{x}_2 + \Omega_1^2 x_2 = -(\Omega_1^2)\left[\alpha - \left(\frac{3A^2}{4}\right)\Omega_1^6 h(\alpha,\beta)\right]\cos\theta + \text{HOH}, \tag{5.2.88}$$

where

$$h(\alpha,\beta)(\alpha^2 - \alpha\beta + 2\beta^2)\alpha. \tag{5.2.89}$$

Setting to zero the coefficient of $\cos\theta$, to prevent the occurrence of a secular term, gives

$$\Omega_1^6 = \left[\left(\frac{4}{3}\right)\frac{1}{A^2}\right]\left[\frac{1}{\alpha^2 - \alpha\beta + 2\beta^2}\right] = \Omega_0^6\left[\frac{1}{\alpha^2 - \alpha\beta + 2\beta^2}\right], \tag{5.2.90}$$

and

$$\Omega_1(A) = \left[\left(\frac{4}{3}\right)^{1/6}\frac{1}{A^{1/3}}\right]\left[\frac{1}{\alpha^2 - \alpha\beta + 2\beta^2}\right]^{1/6} = \frac{1.041424}{A^{1/3}}. \tag{5.2.91}$$

The corresponding percentage error is

$$\left| \frac{\Omega_{\text{exact}} - \Omega_1}{\Omega_{\text{exact}}} \right| \cdot 100 = 2.7\%. \tag{5.2.92}$$

Comparing Eqs. (5.2.83) and (5.2.92), the conclusion is that $\Omega_0(A)$ is a slightly better estimate of the exact value for the angular frequency than $\Omega_1(A)$.

An alternative iteration scheme for the cube-root equation

$$\ddot{x} + x^{1/3} = 0, \tag{5.2.93}$$

is [12]

$$\ddot{x}_{k+1} + \Omega_k^2 x_{k+1} = \Omega_k^2 x_k - x_k^{1/3}. \tag{5.2.94}$$

However, inspection of this formula shows that it can only be applied to obtain one level of iteration. Currently no known expansion exists for

$$f(\theta) = (c_1 \cos\theta + c_2 \cos 3\theta + \cdots)^{1/3}, \tag{5.2.95}$$

where $(c_1, c_2, \dots)$ are the constant coefficients, although theoretical reasoning informs us that [13, 14]

$$f(\theta) = d_1 \cos\theta + d_2 \cos 3\theta + \cdots, \tag{5.2.96}$$

for some set of coefficients, $\{d_i\}$.

For $k = 0$, Eq. (5.2.94) is

$$\ddot{x}_1 + \Omega_0^2 x_1 = \Omega_0^2 x_0 - x_0^{1/3} = \Omega_0^2 A \cos\theta - (A\cos\theta)^{1/3}, \tag{5.2.97}$$

where $\theta = \Omega_0 t$. Now $(\cos\theta)^{1/3}$ has the Fourier expansion

$$(\cos\theta)^{1/3} = \sum_{n=0}^{\infty} a_{2n+1} \cos(2n+1)\theta \tag{5.2.98}$$

where

$$a_{2n+1} = \frac{3\Gamma\left(\frac{7}{3}\right)}{2^{4/3}\Gamma\left(n + \frac{5}{3}\right)\Gamma\left(\frac{2}{3} - n\right)}, \tag{5.2.99}$$

and

$$a_1 = 1.15959526696\dots. \tag{5.2.100}$$

If these results are substituted into the right-hand side of Eq. (5.2.97), then the following result is found

$$\ddot{x}_1 + \Omega_0^2 x_1 = (\Omega_0^2 A - A^{1/3} a_1)\cos\theta - A^{1/3} \sum_{n=1}^{\infty} a_{2n+1} \cos(2n+1)\theta. \tag{5.2.101}$$

The elimination of a secular term in the solution for $x_1(t)$ requires that the coefficient of $\cos\theta$ be zero, i.e.,

$$\Omega_0^2 A - A^{1/2} a_1 = 0$$

or solving for $\Omega_0(A)$,

$$\Omega_0(A) = \frac{\sqrt{a_1}}{A^{1/3}} = \frac{1.076845}{A^{1/3}}. \tag{5.2.102}$$

Therefore, $x_1(t)$ is the solution to the differential equation

$$\ddot{x}_1 + \Omega_0^2 x_1 = -A^{1/3} \sum_{n=1}^{\infty} a_{2n+1} \cos(2n+1)\theta, \tag{5.2.103}$$

where $\theta = \Omega_0 t$ and $x_1(0) = A$ with $\dot{x}(0) = 0$. The full solution for $x_1(t)$ is

$$\begin{aligned} x_1(t) &= \beta A \cos[\Omega_0(A)t] \\ &\quad + A \sum_{n=1}^{\infty} \left\{ \frac{a_{2n+1}}{a_1[(2n+1)^2 - 1]} \right\} \cos[(2n+1)\Omega_0(A)t], \end{aligned} \tag{5.2.104}$$

where $\Omega_0(A)$ is taken from Eq. (5.2.102) and β is the constant [12]

$$\beta = 1 - \sum_{n=1}^{\infty} \frac{a_{2n+1}}{a_1[(2n+1)^2 - 1]}. \tag{5.2.105}$$

Note that the percentage error is

$$\left| \frac{\Omega_{\text{exact}} - \Omega_0}{\Omega_{\text{exact}}} \right| \cdot 100 = 0.6\% \text{ error}. \tag{5.2.106}$$

Therefore, in spite of the limitations of the single-step iteration scheme, given by Eq. (5.2.94), this procedure provides an accurate estimation of the value for the angular frequency. Inspection of Eq. (5.2.104) demonstrates that all harmonics appear in its representation.

5.2.6 $\ddot{x} + x^{-1/3} = 0$

An iteration scheme for the inverse-cubic TNL oscillator is obtained by the following manipulations,

$$1 = -\ddot{x} x^{1/3}$$

$$1 = -(\ddot{x})^3 x$$

$$x = -(\ddot{x})^3 x^2$$

$$\ddot{x} + \Omega^2 x = \ddot{x} - \Omega^2(\ddot{x})^3 x^2,$$

and in this form, the iteration scheme is

$$\ddot{x}_{k+1} + \Omega_k^2 x_{k+1} = \ddot{x}_k - \Omega_k^2(\ddot{x}_k)^3 x_k^2. \tag{5.2.107}$$

For $k = 0$, with $x_0(t) = A\cos\theta = A\cos(\Omega_0 t)$, we have

$$\begin{aligned}\ddot{x}_1 + \Omega_0^2 x_1 &= \ddot{x}_0 - \Omega_0^2(\ddot{x}_0)^3 x_0^2 \\ &= (-\Omega_0^2 A\cos\theta) - \Omega_0^2(-\Omega_0^2 A\cos\theta)^3(A\cos\theta)^2 \\ &= -(\Omega_0^2)\left[1 - \frac{5\Omega_0^6 A^4}{8}\right] A\cos\theta \\ &\quad + \left(\frac{5\Omega_0^8 A^5}{16}\right)\cos 3\theta + \left(\frac{\Omega_0^8 A^5}{16}\right)\cos 5\theta. \end{aligned} \tag{5.2.108}$$

No secular term in the solution for $x_1(t)$ requires that the coefficient of the $\cos\theta$ term be zero and doing this gives

$$1 - \frac{5\Omega_0^6 A^4}{8} = 0$$

or

$$\Omega_0^6 = \left(\frac{8}{5}\right)\frac{1}{A^4}$$

and

$$\Omega_0(A) = \left(\frac{8}{5}\right)^{1/6}\frac{1}{A^{2/3}} = \frac{1.081484}{A^{2/3}}. \tag{5.2.109}$$

Since $\Omega_{\text{exact}}(A)$ is

$$\Omega_{\text{exact}}(A) = \left(\frac{2}{\sqrt{3}}\right)\frac{1}{A^{2/3}} = \frac{1.154700538}{A^{2/3}}, \tag{5.2.110}$$

the percentage error for $\Omega_0(A)$ is

$$\left|\frac{\Omega_{\text{exact}} - \Omega_0}{\Omega_{\text{exact}}}\right| \cdot 100 = 6.3\% \text{ error}. \tag{5.2.111}$$

To calculate $x_1(t)$, the solution to the following differential equation must be found,

$$\ddot{x}_1 + \Omega_0^2 x_1 = \left(\frac{5\Omega_0^8 A^5}{16}\right)\cos 3\theta + \left(\frac{\Omega_0^8 A^5}{16}\right)\cos 5\theta. \tag{5.2.112}$$

Requiring that $x_1(0) = A$ and $\dot{x}_1(0) = 0$, and using $\Omega_0(A)$ from Eq. (5.2.109), the full solution for $x_1(t)$ is found to be the expression

$$\begin{cases} x_1(t) = \left(\frac{16}{15}\right) A\left[\cos\theta - \left(\frac{15}{256}\right)\cos 3\theta - \left(\frac{1}{256}\right)\cos 5\theta\right] \\ \theta = \Omega_0 t = \left(\frac{8}{5}\right)^{1/6}\left[\frac{t}{A^{2/3}}\right]. \end{cases} \tag{5.2.113}$$

Note that the coefficients decrease rapidly, i.e.,

$$\left|\frac{a_1}{a_0}\right| = \frac{15}{256} = 0.0586, \qquad \left|\frac{a_2}{a_1}\right| = \frac{1}{15} = 0.0667. \tag{5.2.114}$$

The equation for $x_2(t)$ is

$$\ddot{x}_2 + \Omega_1^2 x_2 = \ddot{x}_1 - \Omega_1^2(\ddot{x}_1)^3 x_1^2, \tag{5.2.115}$$

and with $x_1(t)$ given by the expression

$$x_1(t) = \left(\frac{16}{15}\right) A \left[\cos(\Omega_1 t) - \left(\frac{15}{256}\right)\cos(3\Omega_1 t) - \left(\frac{1}{256}\right)\cos(5\Omega_1 t)\right], \tag{5.2.116}$$

an easy calculation finds that the right-hand side of Eq. (5.2.115) contains all odd harmonics from θ to 25θ, i.e.,

$$\ddot{x}_2 + \Omega_1^2 x_2 = \sum_{n=0}^{12} b_n(A, \Omega_1^2)\cos(2n+1)\theta, \tag{5.2.117}$$

where $b_n(A, \Omega_1^2)$ are known functions of A and Ω_1^2. While the full solution for this differential equation can be directly found, significant algebraic manipulation is required to obtain the final result.

5.2.7 $\ddot{x} + x + x^{1/3} = 0$

The modified harmonic oscillator TNL oscillator, with a cube-root term, is [16]

$$\ddot{x} + x + x^{1/3} = 0. \tag{5.2.118}$$

Starting with

$$x^{1/3} = -(\ddot{x} + x)$$

$$x = -(\ddot{x} + x)^3$$

$$\ddot{x} + \Omega^2 x = \ddot{x} - \Omega^2(\ddot{x} + x)^3,$$

we take the associated iteration scheme to be

$$\ddot{x}_{k+1} + \Omega_k^2 x_{k+1} = \ddot{x}_k - \Omega_k^2(\ddot{x}_k + x_k)^3. \tag{5.2.119}$$

Therefore, for $k = 0$, we have with $x_0(t) = A\cos\theta$, $\theta = \Omega_0 t$

$$\begin{aligned}\ddot{x}_1 + \Omega_0^2 x_1 &= \ddot{x}_0 - \Omega_0^2(\ddot{x}_0 + x_0)^3 \\ &= \Omega_0^2\left[-1 + (\Omega_0^2 - 1)^3\left(\frac{3A^2}{4}\right)\right] A\cos\theta\end{aligned}$$

$$+\Omega_0^2(\Omega_0^2-1)^3\left(\frac{3A^2}{4}\right)\cos 3\theta. \tag{5.2.120}$$

The no secular term requirement gives

$$-1+(\Omega_0^2-1)^3\left(\frac{3A^2}{4}\right)=0,$$

or

$$\Omega_0^2(A)=1+\left(\frac{4}{3}\right)^{1/3}\left(\frac{1}{A^{2/3}}\right)=1+\frac{1.100642}{A^{2/3}}. \tag{5.2.121}$$

With this information, it follows that $x_1(t)$ satisfies the equation

$$\ddot{x}_1+\Omega_0^2 x_1=\Omega_0^2(\Omega_0^3-1)^2\left(\frac{A^3}{4}\right)\cos 3\theta, \tag{5.2.122}$$

and this equation has the following full solutions for $x_1(t)$,

$$x_1(t)=A\left[\left(\frac{25}{24}\right)\cos\theta-\left(\frac{1}{24}\right)\cos 3\theta\right]. \tag{5.2.123}$$

To obtain this result, the particular solution was taken to be

$$x_1^{(p)}(t)=D\cos 3\theta, \tag{5.2.124}$$

where D is found to be

$$D=-(\Omega_0^2-1)^3\left(\frac{A^3}{32}\right). \tag{5.2.125}$$

However,

$$(\Omega_0^2-1)^3=\left(\frac{4}{3}\right)\frac{1}{A^2},$$

and when this is substituted into Eq. (5.2.125), D takes the value

$$D=-\left(\frac{4}{3}\right)\left(\frac{1}{A^2}\right)\left(\frac{A^3}{32}\right)=-\left(\frac{A}{24}\right). \tag{5.2.126}$$

The full solution for $x_1(t)$ is

$$x_1(t)=C\cos\theta-\left(\frac{A}{24}\right)\cos 3\theta,$$

and for $x_1(0)=A$, then $C=\frac{25A}{24}$, and the result given in Eq. (5.2.123) is derived.

If the calculation at this point is terminated, then

$$\begin{cases} x_1(t)=A\left[\left(\frac{25}{24}\right)\cos(\Omega_0 t)-\left(\frac{1}{24}\right)\cos(3\Omega_0 t)\right], \\ \Omega_0^2(A)=1+\left(\frac{4}{3}\right)\left(\frac{1}{A^{2/3}}\right). \end{cases} \tag{5.2.127}$$

For purposes of comparison, let us now calculate a first-order harmonic balance approximation to the periodic solution of Eq. (5.2.118). The assumed solution is

$$x_{1+p}(t) = A\cos\theta, \quad \theta = \Omega_{HB}t, \tag{5.2.128}$$

and its substitution into Eq. (5.2.118) gives

$$-\Omega_{HB}^2 A\cos\theta + A\cos\theta + (A\cos\theta)^{1/3} \simeq 0,$$

and

$$[(1-\Omega_{HB}^2)A + a_1 A^{1/3}]\cos\theta + \text{HOH} \simeq 0. \tag{5.2.129}$$

To obtain this result, we made use of the expansion of $(\cos\theta)^{1/3}$ stated in Eq. (5.2.98). Setting the coefficient of $\cos\theta$ to zero gives

$$\Omega_{HB}^2(A) = 1 + \frac{a_1}{A^{2/3}}. \tag{5.2.130}$$

Using $a_1 = 1.159595\ldots$, we find

$$\Omega_{HB}^2(A) = 1 + \frac{1.159595}{A^{2/3}}. \tag{5.2.131}$$

Since no known solution exists for Eq. (5.2.118), we can only compare the two expressions for the angular frequencies, $\Omega_0^2(A)$ and $\Omega_{HB}^2(A)$, respectively, from Eqs. (5.2.121) and (5.2.131). Both formulas give similar results with a percentage error difference of about 5.2%. These calculations suggest that a plot of $\Omega^2(A)$ versus A has the general features presented in Figure 5.2.1; in particular

$$A \text{ small} : \Omega^2(A) \sim \frac{C}{A^{2/3}},$$
$$A \text{ large} : \Omega^2(A) \sim 1,$$

where C is a positive constant.

5.3 Worked Examples: Extended Iteration

The formula for extended iteration is given in Eq. (5.1.17). It corresponds to making a linear Taylor series approximation at $x(t) = x_0(t)$ where

$$x_0(t) = A\cos\theta, \quad \theta = \Omega_k t. \tag{5.3.1}$$

Again, note that at the k-th level of iteration the angular frequency Ω is taken to be Ω_k, i.e., the value for Ω in $x_0(t)$ changes with the order of iteration. For extended iteration, $x_1(t)$ has the same mathematical form as that for the direct iteration method. The methods differ only for $k \geq 2$.

To illustrate the use of the extended iteration procedure, two TNL oscillator equations will be studied. For both cases, $x_2(t)$ is determined.

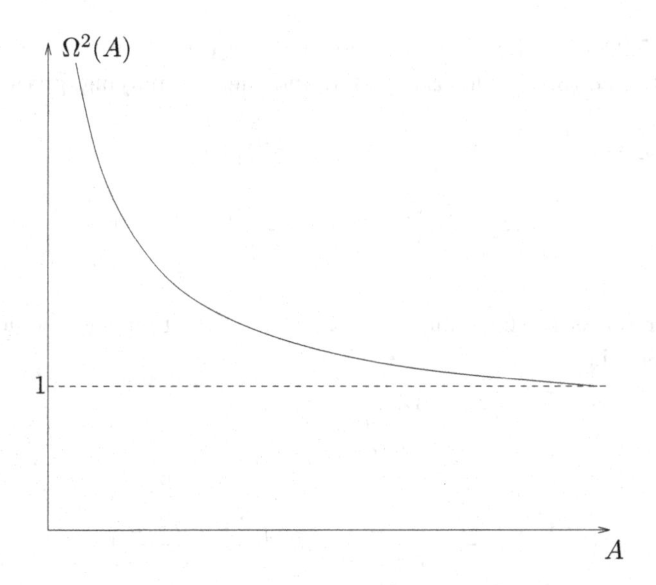

Fig. 5.2.1 Plot of $\Omega^2(A)$ versus A for the periodic solutions of Eq. (5.2.118).

5.3.1 $\ddot{x} + x^3 = 0$

Starting with the differential equation

$$\ddot{x} + x^3 = 0, \tag{5.3.2}$$

we obtain

$$\ddot{x} + \Omega^2 x = \Omega^2 x - x^3 \equiv G(x, \Omega^2), \tag{5.3.3}$$

with

$$G_x(x, \Omega^2) = \Omega^2 - 3x^2. \tag{5.3.4}$$

Therefore, according to Eq. (5.1.17), the related extended iteration scheme is

$$\begin{aligned} \ddot{x}_{k+1} + \Omega_k^2 x_{k+1} &= G(x_0, \Omega_k^2) + G_x(x_0, \Omega_k^2)(x_k - x_0) \\ &= (\Omega_k^2 x_0 - x_0^3) + (\Omega_k^2 - 3x_0^2)(x_k - x_0). \end{aligned} \tag{5.3.5}$$

For $k = 1$, we find

$$\ddot{x}_2 + \Omega_1^2 x_2 = (\Omega_1^2 x_0 - x_0^3) + (\Omega_1^2 - 3x_0^2)(x_1 - x_0), \tag{5.3.6}$$

where

$$\begin{cases} x_0(t) = A\cos\theta, \\ x_1(t) = A[\alpha\cos\theta + \beta\cos 3\theta], \\ \theta = \Omega_1 t, \quad \alpha = \left(\dfrac{23}{24}\right), \quad \beta = \left(\dfrac{1}{24}\right). \end{cases} \tag{5.3.7}$$

(See Eq. (5.2.9) for the above expression for $x_1(t)$.) Substituting the expressions of Eq. (5.3.7) into Eq. (5.3.6) and then simplifying, gives

$$\begin{aligned}\ddot{x}_2 + \Omega_1^2 x_2 &= A\left[\alpha\Omega_1^2 - \left(\frac{3A^2}{4}\right)(3\alpha+\beta-2)\right]\cos\theta \\ &\quad + A\left[\beta\Omega_1^2 + \left(\frac{A^2}{4}\right)(2-\alpha-2\beta)\right]\cos 3\theta - \left(\frac{3\beta A^3}{4}\right)\cos 5\theta.\end{aligned} \tag{5.3.8}$$

No secular terms in the solution for $x_2(t)$ requires that the coefficient of $\cos\theta$ be zero, i.e.,

$$\alpha\Omega_1^2 - \left(\frac{3A^2}{4}\right)(3\alpha+\beta-2) = 0;$$

and

$$\Omega_1^2 = \left(\frac{3A^2}{4}\right)\left(\frac{3\alpha+\beta-2}{\alpha}\right) = \Omega_0^2\left(\frac{3\alpha+\beta-2}{\alpha}\right), \tag{5.3.9}$$

or

$$\Omega_1(A) = (0.846990)A. \tag{5.3.10}$$

Comparing this $\Omega_1(A)$ with the exact value, we obtain the percentage-error

$$\left|\frac{\Omega_{\text{exact}} - \Omega_1}{\Omega_{\text{exact}}}\right| \cdot 100 = 0.03\% \text{ error.} \tag{5.3.11}$$

With the above value for $\Omega_1(A)$, the coefficients of $\cos 3\theta$ and $\cos 5\theta$ can be evaluated and we find

$$\ddot{x}_2 + \Omega_1^2 x_2 = \left(\frac{595}{2208}\right)A^3\cos 3\theta - \left(\frac{69}{2208}\right)A^3\cos 5\theta. \tag{5.3.12}$$

The particular solution for this differential equation is

$$x_2^{(p)}(t) = D_1\cos 3\theta + D_2\cos 5\theta, \tag{5.3.13}$$

and D_1 and D_2 are found to have the values

$$D_1 = \left(\frac{595A^3}{2208}\right)\left(\frac{1}{-8\Omega_1^2}\right) = -\left(\frac{595}{12,672}\right)A,$$

$$D_2 = -\left(\frac{69A^3}{2208}\right)\left(\frac{1}{-24\Omega_1^2}\right) = \left(\frac{23}{12,672}\right)A.$$

Therefore, the full solution is

$$x_2(t) = C\cos\theta + D_1\cos 3\theta + D_2\cos 5\theta,$$

with

$$C = \left(\frac{13,244}{12,672}\right) A,$$

and, finally,

$$x_2(t) = A\left\{\left(\frac{13,244}{12,672}\right)\cos\theta - \left(\frac{595}{12,672}\right)\cos 3\theta + \left(\frac{23}{12,672}\right)\cos 3\theta\right\}, \tag{5.3.14}$$

$$\theta = \Omega_1(A)t = \sqrt{\frac{66}{92}}\,A. \tag{5.3.15}$$

A calculation of the ratio of the coefficients gives

$$\left|\frac{a_1}{a_0}\right| = \frac{595}{13,244} \approx (4.5)\cdot 10^{-2},$$

$$\left|\frac{a_2}{a_1}\right| = \frac{23}{595} \approx (3.9)\cdot 10^{-2}.$$

These results suggest that in a higher level iteration calculation the coefficients of the harmonic trigonometric terms should decrease rapidly.

In summary, the extended iteration procedure gives a more accurate solution in comparison to the direct iteration method.

5.3.2 $\ddot{x} + x^{-1} = 0$

This TNL oscillator has several possible iteration schemes. We use the one derived from the relation

$$\ddot{x} + \Omega^2 x = \Omega^2 x - x(\ddot{x})^2 = G(x, \ddot{x}, \Omega^2), \tag{5.3.16}$$

that is

$$\ddot{x}_{k+1} + \Omega_k^2 x_{k+1} = [\Omega_k^2 x_0 - x_0(\ddot{x}_0)^2] + [\Omega_k^2 - (\ddot{x}_0)^2](x_k - x_0) - 2x_0\ddot{x}_1(\ddot{x}_k - \ddot{x}_0). \tag{5.3.17}$$

To obtain this relation the following formula was used for the extended iteration scheme

$$\ddot{x}_{k+1}\Omega_k^2 x_{k+1} = G(x_0, \ddot{x}_0, \Omega_k^2) + G_x(x_0, \ddot{x}_0, \Omega_k^2)(x_k - x_0) + G_{\ddot{x}}(x_0, \ddot{x}_0, \Omega_k^2)(\ddot{x}_k - \ddot{x}_0).$$

For $k = 1$, we have

$$\ddot{x}_2 + \Omega_1^2 x_2 = 2x_0(\ddot{x}_0)^2 + [\Omega_1^2 - (\ddot{x}_0)^2]x_1 - 2x_0\ddot{x}_0\ddot{x}_1, \tag{5.3.18}$$

with

$$\begin{cases} x_0(t) = A\cos\theta, \\ x_1(t) = A[\alpha\cos\theta + \beta\cos 3\theta], \\ \theta = \Omega_1 t, \quad \alpha = \dfrac{23}{24}, \quad \beta = \dfrac{1}{24}. \end{cases} \tag{5.3.19}$$

(See Eq. (5.2.45) for $x_1(t)$.) Substitution of the items in Eq. (5.3.19) into the right-hand side of Eq. (5.3.18) gives, after some algebraic and trigonometric simplification, the result

$$\begin{aligned} \ddot{x}_2 + \Omega_1^2 x_2 = (\Omega_1^2 A)&\left[\alpha - (3 - 7\beta)\left(\frac{\Omega_1^2 A^4}{4}\right)\right]\cos\theta \\ &- \left(\frac{A\Omega_1^2}{4}\right)[(1 + 35\beta)\Omega_1^2 A^2 - 4\beta]\cos 3\theta \\ &- \left(\frac{19\beta}{4}\right)(\Omega_1^4 A^3)\cos 5\theta. \end{aligned} \tag{5.3.20}$$

Setting the coefficient of $\cos\theta$ to zero and solving for Ω_1^2 gives

$$\Omega_1^2(A) = \left[\left(\frac{4}{3}\right)\frac{1}{A^2}\right]\left(\frac{69}{65}\right) = \Omega_0^2(A)\left[\frac{69}{65}\right], \tag{5.3.21}$$

or

$$\Omega_1(A) = \frac{1.189699}{A}. \tag{5.3.22}$$

Comparing $\Omega_1(A)$ with the exact value, $\Omega_{\text{exact}}(A)$, we find the following percentage error

$$\left|\frac{\Omega_{\text{exact}} - \Omega_1}{\Omega_{\text{exact}}}\right| \cdot 100 = 5.1\% \text{ error}. \tag{5.3.23}$$

Note that using the direct iteration scheme, we found

$$\Omega_0(A) = \frac{1.1547}{A} \quad (7.9\% \text{ error}),$$

$$\Omega_1(0) = \frac{1.0175}{A} \quad (18.1\% \text{ error}).$$

Therefore, the extended iteration procedure provides a better estimate of the angular frequency.

Replacing $\Omega_1^2 A^2$ in Eq. (5.3.20), by the expression of Eq. (5.3.21), we obtain

$$\ddot{x}_2 + \Omega_1^2 x_2 = -\left(\frac{A\Omega_1^2}{4}\right)\left(\frac{1292}{390}\right)\cos 3\theta - \left(\frac{A\Omega_1^2}{4}\right)\left(\frac{437}{390}\right)\cos 5\theta. \tag{5.3.24}$$

The corresponding particular solution takes the form

$$x_2^{(p)}(t) = D_1 \cos 3\theta + D_1 \cos 7\theta.$$

Substituting this into Eq. (5.3.25) and equating the coefficients, respectively, of the $\cos 3\theta$ and $\cos 7\theta$ terms, allows the calculation of D_1 and D_2; they are

$$D_1 = \left(\frac{3876}{37,440}\right) A, \quad D_2 = \left(\frac{437}{37,440}\right) A.$$

Since the full solution for $x_2(t)$ is

$$x_2(t) = C \cos\theta + x_1^{(p)}(0)$$

with $x_2(0) = A$, it follows that

$$C = A - D_1 - D_2 = \left(\frac{33,127}{37,440}\right) A,$$

and

$$\begin{cases} x_2(t) = A\left[\left(\frac{33,127}{37,440}\right)\cos\theta + \left(\frac{3876}{37,440}\right)\cos 3\theta + \left(\frac{437}{37,440}\right)\cos 5\theta\right], \\ \theta = \Omega_1(t)t = \left[\frac{92}{65}\right]^{1/2}\left(\frac{1}{A}\right). \end{cases} \tag{5.3.25}$$

Inspection of $x_2(t)$ indicates that the coefficients of the harmonics satisfy the ratios

$$\frac{a_1}{a_0} = \frac{3876}{33,127} \approx 0.117,$$
$$\frac{a_2}{a_1} = \frac{437}{3876} \approx 0.113.$$

5.4 Discussion

The rewriting of a TNL differential equation to a new form raises several mathematical issues. The most significant is the relationship between the solutions of the original equations and those of the reformulated equation. This is illustrated by the equation

$$\ddot{x} + x^{1/3} = 0. \tag{5.4.1}$$

Writing it as

$$\ddot{x} = -x^{1/3},$$

and cubing both sides gives

$$(\ddot{x})^3 + x = 0. \tag{5.4.2}$$

However, the last equation can be factored as follows

$$(\ddot{x} + x^{1/3})\left[\ddot{x} - \left(\frac{1+\sqrt{3}\,i}{2}\right)x^{1/3}\right]\left[\ddot{x} - \left(\frac{1-\sqrt{3}\,i}{2}\right)x^{1/3}\right] = 0, \tag{5.4.3}$$

and this expression corresponds to three nonlinear, second-order differential equations

$$\begin{cases} \ddot{x} + x^{1/3} = 0, \\ \ddot{x} - \left(\dfrac{1+\sqrt{3}\,i}{2}\right)x^{1/3} = 0, \\ \ddot{x} - \left(\dfrac{1-\sqrt{3}\,i}{2}\right)x^{1/3} = 0. \end{cases} \tag{5.4.4}$$

The first of these differential equations has real solutions, while the other two have complex valued solutions. Thus, the original TNL oscillator equation (5.4.1) and its reformulation, Eq. (5.4.2), are not identical in terms of possible solutions. This fact may have an influence on the accuracy of the approximations to the periodic solutions and their angular frequencies.

Re-examination of parameter expansion methods, given in Chapter 4, indicates a similarity with the iteration procedures of this chapter. For lower orders of calculation, the two techniques have many common features such as the differential equations to be solved and, as a consequence, the same solutions and predicted values for the angular frequencies. An interesting research problem would involve investigating possible mathematical connections between these two techniques.

The next section presents a brief summary of several of the advantages and difficulties of iteration methods.

5.4.1 *Advantages of Iteration Methods*

- Only linear, inhomogeneous differential equations are required to be solved at each level of the calculation.
- In principle, iteration methods may be generalized to higher-order differential equations. An important class of such equations are the nonlinear "jerk equations" [17–19]. A particular example is [17]

$$\dddot{x} + a\dot{x} + b\dot{x}^3 + cx^2\dot{x} + dx\dot{x}\ddot{x} + e\dot{x}(\ddot{x})^2 = 0,$$

  where (a, b, c, d, e) are constants.

- The coefficients of the higher harmonics, for a given value of the iteration index k, decrease rapidly with increasing harmonic number. This implies that higher-order (in k) solutions may not be required.
- The extended iteration method generally is easier to apply, for a given equation, in comparison with similar direct iteration techniques, i.e., it requires fewer overall computations for the calculation of $x(t)$ and $\Omega(k)$ for a given value of k. In particular, for equations having cubic-type nonlinearities, the number of harmonics at the k-th level has approximately the following behaviors

$$\text{direct iteration}: \frac{3^k+1}{2},$$
$$\text{extended iteration}: k+1.$$

Since the coefficients of the harmonic have a rapid decrease in values, the extended iteration method is expected to be sufficient for most investigations.

5.4.2 *Disadvantages of Iteration Methods*

- A given TNL oscillator equation may have more than one possible iteration scheme. At present, there are no *a priori* meta-principles which place limitations on the construction of iteration schemes.
- The first-order calculation of the angular frequency may be more accurate than values calculated using a higher stage of iteration.
- For level $k \geq 2$ calculations, the work required to determine the angular frequency and associated periodic solution may become algebraically intensive.
- Iteration methods may not provide accurate values for the angular frequencies when the original TNL oscillator differential equations contains "singular terms." For example, the equation

$$\ddot{x} + \frac{1}{x} = 0,$$

has the singular term x^{-1}, i.e., it is not defined at $x = 0$. While the solution $x(t)$ and its first derivative exist at $x = 0$, the differential equation is not defined for this value of x. Another example is

$$\ddot{x} + \frac{1}{x^{1/3}} = 0.$$

In this case, the singularity is integrable and the iteration procedures give good results for the periodic solutions.

Problems

5.1 Give reasons why $x_0(t)$ should be selected in the form expressed by Eq. (5.1.5).

5.2 How would the iteration procedure be altered if initial conditions were changed to

$$x_{k+1}(0) = A, \quad \dot{x}_{k+1}(0) = B, \quad k = 0, 1, 2, \ldots ?$$

5.3 Derive an iteration scheme involving higher-order terms in the Taylor series expansion of $G(\ddot{x}, \dot{x}, x)$. See Eqs. (5.1.9) and (5.1.10). What are the advantages (if any) and limitations of such a generalization?

5.4 Calculate the f_i $(i = 1, 2, \ldots, 5)$ listed in Eq. (5.2.14).

5.5 Complete the details and determine the coefficients (D_1, D_2, D_3, D_4) for the particular solution of Eq. (5.2.24).

5.6 Construct a second iteration scheme for

$$\ddot{x} + \text{sgn}(x) = 0.$$

See Section 5.2.4.

5.7 Is there a fundamental difference between the two representations

$$\ddot{x} + x^{1/3} = 0,$$
$$\ddot{x} + |x|^{1/3}\text{sgn}(x) = 0?$$

5.8 Calculate the result for $x_2(t)$ as expressed by Eq. (5.2.88).

5.9 Discuss the derivation of Eq. (5.2.94) and explain why it cannot be extended to $k \geq 1$.

5.10 Derive a second iteration scheme for

$$\ddot{x} + x^{-1/3} = 0.$$

See Section 5.2.6.

5.11 Carry out the steps required to determine $x_2(t)$ and $\Omega_1(A)$ as given by Eqs. (5.3.14) and (5.3.15) for the oscillator

$$\ddot{x} + x^3 = 0.$$

5.12 Provide a possible explanation as to why the ratio of the coefficients for $x_2(t)$ are larger for

$$\ddot{x} + x^{-1} = 0$$

than for

$$\ddot{x} + x^3 = 0.$$

See Sections 5.3.1 and 5.3.2.

References

[1] R. E. Mickens, *Journal of Sound and Vibration* **116**, 185 (1987).
[2] C. W. Lim and B. S. Wu, *Journal of Sound and Vibration* **257**, 202 (2002).
[3] R. E. Mickens, *Journal of Sound and Vibration* **287**, 1045 (2005).
[4] A. H. Nayfeh, *Perturbation Methods* (Wiley, New York, 1973).
[5] R. E. Mickens, *Nonlinear Oscillations* (Cambridge University Press, New York, 1991).
[6] R. E. Mickens, *Journal of Sound and Vibration* **258**, 398 (2002).
[7] A. E. Taylor and W. R. Mann, *Advanced Calculus* (Wiley, New York, 1983).
[8] H. Hu, *Journal of Sound and Vibration* **298**, 446 (2006).
[9] R. E. Mickens, *Oscillations in Planar Dynamic Systems* (World Scientific, Singapore, 1996).
[10] R. E. Mickens, *Journal of Sound and Vibration* **306**, 968 (2007).
[11] T. Lipscomb and R. E. Mickens, *Journal of Sound and Vibration* **169**, 138 (1994).
[12] R. E. Mickens, *Journal of Sound and Vibration* **292**, 964 (2006).
[13] H. S. Carslaw, *Introduction to the Theory of Fourier Series and Integrals*, 3rd. ed. (Dover, New York, 1952).
[14] T. W. Körner, *Fourier Analysis* (Cambridge University Press, Cambridge, 1988).
[15] R. E. Mickens, *Mathematical Methods for the Natural and Engineering Sciences* (World Scientific, Singapore, 2004). See Section 2.7.
[16] R. E. Mickens and D. Wilkerson, *Advances in Applied Mathematics and Mechanics* **1**, 383 (2009).
[17] H. P. W. Gottlieb, *Journal of Sound and Vibration* **271**, 671 (2004).
[18] B. S. Wu, C. W. Lim, and W. P. Sun, *Physics Letters A* **354**, 95 (2006).
[19] H. Hu, *Physics Letters A* **372**, 4205 (2008).

Chapter 6

Averaging Methods

All of the previous methods for calculating approximations to the periodic solutions have one outstanding limitation: they cannot be applied to TNL differential equations having transitory solutions, i.e., the solutions may be oscillatory, but not periodic, or the equations may have limit cycles, with transitory behavior for nearby solutions [1, 2]. In general, these systems contain dissipation and this causes the "amplitude" and the "phase" of the oscillations to change with time.

A technique to resolve these issues is the method of averaging [1–3]. The basic procedure begins with the assumption that the oscillatory solution can be written as

$$x(t) = a(t)\cos\psi(t).$$

Next, exact first-order differential equations are derived for $a(t)$ and $\psi(t)$. However, these equations are of such complexity that they cannot be solved in closed form. The application of a suitable "averaging" leads to two other first-order approximate equations for the amplitude, $a(t)$, and the phase $\psi(t)$. In general, these latter two differential equations can be solved exactly. If we denote the respective solutions by $\bar{a}(t)$ and $\bar{\psi}(t)$, then the approximation to the solution $x(t)$ is

$$x(t) \simeq \bar{x}(t) = \bar{a}(t)\cos\bar{\psi}(t).$$

This chapter presents several averaging procedures for determining approximations to the solution of TNL oscillatory differential equations. The significance and meaning of the term "averaging methods" can be explained by the manner in which the various methods are derived.

In Section 6.1, two elementary TNL averaging methods will be presented. These procedures are based on early work by Mickens and Oyedeji [4], and Mickens [5]. Section 6.2 gives a number of worked examples based on these two procedures. In Section 6.3, we present a new analytical method derived by Cveticanin [5] for solving TNL oscillator equations containing terms corresponding to dissipative. Section 6.4 contains several worked examples based on the Cveticanin method. Section 6.5 gives a brief reference and discussion of related work by other researchers. Finally, in Section 6.6, we comment on the advantages and disadvantages of these various averaging procedures.

6.1 Elementary TNL Averaging Methods

6.1.1 *Mickens-Oyedeji Procedure*

Consider the following special TNL oscillator differential equation

$$\ddot{x} + x^3 = \epsilon F(x, \dot{x}), \quad 0 < \epsilon \ll 1, \tag{6.1.1}$$

where ϵ is a small parameter and F is a polynomial function of x and $\dot{x}$. Assume that the solution to this differential equation takes the form

$$x(t) = a(t)\cos[\Omega t + \phi(t)], \tag{6.1.2}$$

where, for the moment, the dependency on ϵ is suppressed. The functions $a(t)$ and $\phi(t)$ are unknown functions, and they and the unknown constant Ω must be determined. The quantities $a(t)$ and $\psi(t) = \Omega t + \phi(t)$ are, respectively, the amplitude and the phase of the oscillation.

Taking the derivative of Eq. (6.1.2) gives

$$\dot{x} = -\Omega a \sin\psi + \dot{a}\cos\psi - a\dot{\phi}\sin\psi. \tag{6.1.3}$$

If we require that

$$\dot{x} = -\Omega a \sin\psi, \tag{6.1.4}$$

then it follows that

$$\dot{a}\cos\psi - a\dot{\phi}\sin\psi = 0, \tag{6.1.5}$$

and the second derivative of x is

$$\ddot{x} = -\Omega\dot{a}\sin\psi - \Omega a\dot{\phi}\cos\psi - \Omega^2 a\cos\psi. \tag{6.1.6}$$

If Eqs. (6.1.2), (6.1.4) and (6.1.6) are substituted into Eq. (6.1.1), then we find

$$\dot{a}\sin\psi + A\dot{\phi}\cos\psi = -\Omega a\cos\psi + \left(\frac{3a^3}{4\Omega}\right)\cos\psi$$

$$+ \left(\frac{a^3}{4\Omega}\right) \cos 3\psi - \left(\frac{\epsilon}{\Omega}\right) F(a\cos\psi, -\Omega a \sin\psi). \tag{6.1.7}$$

Equations (6.1.5) and (6.1.7) are linear in $\dot{a}$ and $\dot{\phi}$, and solving them for these quantities gives

$$\dot{a} = -\,\Omega a \cos\psi \sin\psi + \left(\frac{3a^3}{4\Omega}\right) \cos\psi \sin\psi$$
$$+ \left(\frac{a^3}{4\Omega}\right) \cos 3\psi \sin\psi - \left(\frac{\epsilon}{\Omega}\right) F \sin\psi, \tag{6.1.8}$$

$$a\dot{\phi} = \left(\frac{3a^3}{4\Omega} - \Omega a\right)(\cos\psi)^2 + \left(\frac{a^3}{4\Omega}\right) \cos 3\psi \cos\psi - \left(\frac{\epsilon}{\Omega}\right) F \cos\psi, \tag{6.1.9}$$

where $F = F(a\cos\psi, -a\Omega\sin\psi)$. Note that these equations are expressions for $\dot{a}$ and $\dot{\phi}$, but, in general, cannot be solved for $a(t)$ and $\phi(t)$. However, approximate formulas can be derived by making use of the fact that the right-sides of these equations are both periodic in ψ with period 2π. Therefore, averaging the right-sides over 2π gives

$$\dot{a} = -\left(\frac{\epsilon}{2\pi\Omega}\right) \int_0^{2\pi} F(a\cos\psi, -\Omega a\sin\psi) \sin\psi \, d\psi, \tag{6.1.10}$$

$$\dot{\phi} = -\left(\frac{\epsilon}{2\pi\Omega a}\right) \int_0^{2\pi} F(a\cos\psi, -\Omega a\sin\psi) \cos\psi \, d\psi + \left(\frac{1}{2}\right)\left(\frac{3a^2}{4\Omega} - \Omega\right). \tag{6.1.11}$$

Strictly speaking, in the last two equations, the (a, ϕ) functions should be represented by notation such as $(\bar{a}, \bar{\phi})$ to indicate that they are averaged quantities. However, in keeping with the usual practice, no such over-bars will be used.

In summary the Mickens-Oyedeji [4] generalization of the method of first-order averaging [1, 2] applied to the equation

$$\ddot{x} + x^3 = \epsilon F(x, \dot{x}), \quad 0 < \epsilon \ll 1,$$

is

$$x(t) \simeq a(t) \cos\psi(t),$$

where $a(t)$ and $\phi(t)$ are determined from solving respectively, Eqs. (6.1.10) and (6.1.11).

The above presentation applies only to the special case of TNL oscillator equations where the TNL "elastic force" term is cubic. Also, the derivation is based on the use of trigonometric functions, i.e., $\sin\psi$ and $\cos\psi$. However, in a series of publications [8], Bejarano and Yuste, have extended this methodology for Eq. (6.1.1) to the use of Jacobi elliptic functions.

Finally, it should be observed that the constant Ω is not specified. The worked examples in Section 6.2 will give one possibility for making this selection.

6.1.2 *Combined Linearization and Averaging Method*

Consider the equation

$$\ddot{x} + g(x) = \epsilon F(x, \dot{x}), \quad 0 < \epsilon \ll 1, \tag{6.1.12}$$

where $g(x)$ is a nonlinear function. The basic idea of the combined linearization and averaging (CLA) method [5, 7] is to replace $g(x)$ by an appropriate linear approximation, i.e.,

$$g(x) \to \Omega^2 x, \tag{6.1.13}$$

such that the resulting replacement equation

$$\ddot{x} + \Omega^2 x = \epsilon F(x, \dot{x}), \quad 0 < \epsilon \ll 1, \tag{6.1.14}$$

can be solved by means of any of the standard perturbation procedures [1–3, 9]. Thus, for this method the fundamental issue is how to determine Ω^2 and what parameters it depends on; for example, it may be dependent on the initial conditions, as well as parameters appearing in the original differential equation. The general goal of this procedure is to determine a solution to Eq. (6.1.14) that is "close" to the actual solution of Eq. (6.1.12). In the following work, only approximations to $O(\epsilon)$ are given. In any case, whether or not the solutions of Eqs. (6.1.12) and (6.1.14) are "close," we do expect that they will have many of the same general qualitative properties.

Note that the linearization of a function is an ambiguous task and just how this should be done is dependent not only on the particular problem under consideration, but also on the exact purposes such a linearization is to accomplish. Two references giving a broad range of linearization procedures are the works of Bellman [10] and Zwillinger [11].

Our method of linearization is "harmonic linearization," i.e., replace x by

$$x \to a\cos\theta, \tag{6.1.15}$$

substitute this into $g(x)$ to obtain

$$g(x) \to g(a\cos\theta) = \sum_{k=0}^{\infty} g_k(a)\cos(2k+1)\theta, \tag{6.1.16}$$

where the coefficients may be determined as functions of a and use the replacement by the first term, i.e.,

$$g(x) \to g_0(a)\cos\theta = \left[\frac{g_0(a)}{a}\right](a\cos\theta) \equiv \Omega^2(a)x. \tag{6.1.17}$$

This linearization may be applied for both TNL and standard oscillators.

The following three examples illustrate the harmonic linearization procedure.

First, consider $g(x) = x^3$. Following the steps given above, we find

$$\begin{aligned} g(x) = x^3 \to (a\cos\theta)^3 &= \left(\frac{3a^3}{4}\right)\cos\theta + \left(\frac{a^3}{4}\right)\cos 3\theta \\ &= \left(\frac{3a^2}{4}\right)(a\cos\theta) + \text{HOH}, \end{aligned}$$

and

$$g(x) = x^3 \to \Omega^2(a)x, \quad \Omega^2(a) = \frac{3a^2}{4}. \tag{6.1.18}$$

Second, for the function $g(x) = x + \lambda x^3$, we have

$$\begin{aligned} g(x) \to (a\cos\theta) + \lambda(a\cos\theta)^3 &= (a\cos\theta) + \left(\frac{3\lambda a^3}{4}\right)\cos\theta + \text{HOH} \\ &= \left[1 + \frac{3\lambda a^2}{4}\right](a\cos\theta) + \text{HOH} = \Omega^2(a)x, \end{aligned}$$

where

$$\Omega^2(a) = 1 + \frac{3\lambda a^2}{4}. \tag{6.1.19}$$

Third, for $g(x) = x^{1/3}$, we find

$$g(x) = x^{1/3} \to (a\cos\theta)^{1/3} = g_0(a)\cos\theta + \text{HOH} \tag{6.1.20}$$

where

$$g_0(a) = (1.1596)a^{1/3},$$

and

$$\Omega^2(a) = \frac{1.1596}{a^{2/3}}. \tag{6.1.21}$$

Therefore the TNL oscillator differential equation

$$\ddot{x} + x^{1/3} = \epsilon(1 - x^2)\dot{x},$$

becomes, under harmonic linearization, the equation

$$\ddot{x} + \left(\frac{1.1596}{a^{2/3}}\right)x = \epsilon(1 - x^2)\dot{x}.$$

If $F(x,\dot{x})$ is a function only of x, then the system is a conservative oscillator and any of the previous methods, covered in Chapters 3, 4, and 5, may be used to calculate an approximation to the periodic solutions. The more interesting case arises when $F(x,\dot{x})$ depends on both x and $\dot{x}$, and for this situation limit-cycles may occur [1–3]. For this situation, the harmonic linearized equation

$$\ddot{x}+\Omega^2(a)x=\epsilon F(x,\dot{x}), \quad 0<\epsilon\ll 1, \tag{6.1.22}$$

can be solved by a first-order averaging technique [1–3]. The procedure requires the following steps

(i) Begin with Eq. (6.1.22) and replaced a by $\bar{A}$, an unspecified constant, i.e.,

$$\ddot{x}+\Omega^2(\bar{A})x=\epsilon F(x,\dot{x}). \tag{6.1.23}$$

(ii) Apply standard first-order (in ϵ) averaging to obtain the solution

$$x(t,\epsilon)=a(t,\epsilon)\cos[\Omega(\bar{A})t+\phi(t,\epsilon)], \tag{6.1.24}$$

where

$$\begin{cases} \dot{a}=-\left[\dfrac{\epsilon}{2\pi\Omega(\bar{A})}\right]\displaystyle\int_0^{2\pi} F(a\cos\psi,-\Omega(\bar{A})a\sin\psi)\sin\psi\, d\psi, \\ \dot{\phi}=-\left[\dfrac{\epsilon}{2\pi\Omega(\bar{A})a}\right]\displaystyle\int_0^{2\pi} F(a\cos\psi,-\Omega(\bar{A})a\sin\psi)\cos\psi\, d\psi. \end{cases} \tag{6.1.25}$$

Note that these two equations have the forms

$$\begin{cases} \dot{a}=-\left[\dfrac{\epsilon}{2\pi\Omega(\bar{A})}\right]H_1(a), \\ \dot{\phi}=-\left[\dfrac{\epsilon}{2\pi\Omega(\bar{A})}\right]H_2(a), \end{cases} \tag{6.1.26}$$

where $H_1(a)$ and $H_2(a)$ may be found by comparing Eqs. (6.1.25) and (6.1.26), and, in general, $H_1(0)=0$.

(iii) If $H_1(a)>0$ for $a>0$, then Eq. (6.1.23) is purely dissipative and select $\bar{A}$ to be A, where the initial conditions are $x(0)=A$ and $\dot{x}(0)=0$.

(iv) If $H_1(a)=0$ has a unique, positive zero, i.e., $a=A^*$, then select $\bar{A}$ to be

$$\bar{A}=A^*, \tag{6.1.27}$$

and we have

$$\Omega^2(\bar{A})=\Omega^2(A^*)=\frac{g_0(A^*)}{A^*}. \tag{6.1.28}$$

(v) For the conditions of either (iii) or (iv), solve Eqs. (6.1.26) for $a(t,\epsilon,A^*)$ and $\phi(t,\epsilon A^*)$, subject to the conditions

$$a(0,\epsilon,A^*) = A, \quad \phi(0,\epsilon,A^*) = 0. \tag{6.1.29}$$

In general, the requirements of Eq. (6.1.29) will give for the approximate solution only

$$x(0) = A, \quad \dot{x}(0) = O(\epsilon), \tag{6.1.30}$$

and not $x(0) = A$, $\dot{x}(0) = 0$. This result is a consequence of the first-order averaging procedure.

(vi) Finally, this method gives the following approximation to the periodic solutions of

$$\ddot{x} + g(x) = \epsilon F(x,\dot{x}), \quad 0 < \epsilon \ll 1;$$

$$x(t) \simeq a(t,\epsilon,A^*)\cos\psi(t,\epsilon,A^*),$$

$$\psi(t,\epsilon,A^*) = \Omega(A^*)t + \phi(t,\epsilon,A^*),$$

where $a(t,\epsilon,A^*)$ and $\phi(t,\epsilon,A^*)$ are solutions to Eqs. (6.1.26) and (6.1.29).

This procedure gives not only the limit-cycle parameters,

$$\begin{cases} \text{amplitude} = A^*, \text{ from } H_1(A^*) = 0, \\ \text{frequency} = \Omega^2(A^*) = \dfrac{g_0(A^*)}{A^*}, \end{cases}$$

but also allows the transient motion to be (approximately) determined.

The next section illustrates the application of these two methods.

6.2 Worked Examples

6.2.1 $\ddot{x} + x^3 = -2\epsilon\dot{x}$

This equation corresponds to a linear damped Duffing equation and for this case

$$F(x,\dot{x}) = -2\dot{x}, \tag{6.2.1}$$

with

$$F \rightarrow (-2)(-\Omega a\sin\psi) = 2\Omega a\sin\psi. \tag{6.2.2}$$

Therefore, according to the Mickens-Oyedeji procedure

$$\dot{a} = -\left(\frac{\epsilon}{2\pi\Omega}\right)\int_0^{2\pi}(2\Omega a\sin\psi)\sin\psi\, d\psi = -\epsilon a, \tag{6.2.3}$$

and

$$\begin{aligned}\dot{\phi} &= -\left(\frac{\epsilon}{2\pi\Omega a}\right)\int_0^{2\pi}(2\Omega a)\sin\psi\cos\psi\, d\psi + \left(\frac{1}{2}\right)\left[\frac{3a^2}{4\Omega} - \Omega\right] \\ &= \left(\frac{1}{2\Omega}\right)\left[\frac{3a^2}{4} - \Omega^2\right]. \end{aligned} \tag{6.2.4}$$

The solution to (6.1.3) is

$$a(t, \epsilon) = Ae^{-\epsilon t}. \tag{6.2.5}$$

If we take Ω^2 to be $\frac{3A^2}{4}$, then

$$\dot{\phi} = \left(\frac{1}{2\Omega}\right)\left[\frac{3a^2}{4} - \Omega^2\right] = \left[\left(\frac{\sqrt{3}}{4}\right)A\right]e^{-2\epsilon t} - \left(\frac{\sqrt{3}}{4}\right)A, \tag{6.2.6}$$

and this equation has the solution

$$\phi(t, \epsilon) = \left(\frac{\sqrt{3}A}{4}\right)\left(\frac{1}{2\epsilon}\right)\left[1 - e^{-2\epsilon t}\right] - \left(\frac{\sqrt{3}}{4}\right)At. \tag{6.2.7}$$

To obtain this result, the condition $\phi(0, \epsilon) = 0$ was imposed on the solution for $\phi(t, \epsilon)$.

Since $\psi(t, \epsilon) = \Omega t + \phi(t, \epsilon)$, it follows that

$$\begin{aligned}\psi(t, \epsilon) &= \left(\frac{\sqrt{3}}{2}\right)At + \left(\frac{\sqrt{3}A}{4}\right)\left(\frac{1}{2\epsilon}\right)\left[1 - e^{-2\epsilon t}\right] - \left(\frac{\sqrt{3}}{4}\right)At \\ &= \left(\frac{\sqrt{3}}{4}\right)At + \left[\left(\frac{\sqrt{3}}{A}\right)A\right]\left[\frac{1 - e^{-2\epsilon t}}{2\epsilon}\right]. \end{aligned} \tag{6.2.8}$$

Note that

$$\underset{t-\text{small}}{\psi(t, \epsilon) \simeq} \left(\frac{\sqrt{3}}{2}\right)At, \quad 0 < \epsilon \ll 1;$$

$$\underset{\epsilon \to 0}{\psi(t, \epsilon) =} \left(\frac{\sqrt{3}}{2}\right)At.$$

Thus, given the nature of the approximation procedure, our results are consistent with the properties of previous calculations.

Finally, we have for

$$\ddot{x} + x^3 = -2\epsilon\dot{x},$$

the approximate solution

$$x(t, \epsilon) \simeq Ae^{-\epsilon t}\cos\left\{\left(\frac{\sqrt{3}}{4}\right)At + \left(\frac{\sqrt{3}}{4}\right)A\left[\frac{1 - e^{-2\epsilon t}}{2\epsilon}\right]\right\}. \tag{6.2.9}$$

6.2.2 $\ddot{x} + x^3 = -\epsilon\dot{x}^3$

We now examine the Mickens-Oyedeji solution for the cubic-damped Duffing equation. For this case

$$F(x,\dot{x}) = -(\dot{x})^3 \to -(-\Omega a \sin\psi)^3$$

and

$$F(a\cos\psi, -\Omega a\sin\psi) = \Omega^3 a^3 (\sin\psi)^3 = \Omega^3 a^3 \left[\left(\frac{3}{4}\right)\sin\psi + \left(\frac{1}{4}\right)\sin 3\psi\right].$$

Therefore,

$$\dot{a} = -\left(\frac{3\epsilon}{8}\right)\Omega^2 a^3, \tag{6.2.10}$$

$$\dot{\phi} = -\left(\frac{1}{2\Omega}\right)\left[\frac{3a^2}{4} - \Omega^2\right]. \tag{6.2.11}$$

If Ω^2 is selected to be $3A^2/4$, then

$$\dot{a} = -\left(\frac{9\epsilon A^2}{32}\right)a^3, \tag{6.2.12}$$

and this differential equation has the solution

$$a(t,\epsilon) = \frac{A}{\left[1 + \epsilon\left(\frac{9A^4}{16}\right)t\right]^{1/2}}. \tag{6.2.13}$$

To obtain this result, we used $a(0,\epsilon) = A$.

Substitution of Eq. (6.2.13) into Eq. (6.2.11) and using $\Omega^2 = 3A^2/4$, gives

$$\dot{\phi} = \left(\frac{\sqrt{3}A}{4}\right)\left[\left(\frac{1}{1+\beta t}\right) - 1\right] = -\left(\frac{\sqrt{3}A}{4}\right)\left(\frac{\beta t}{1+\beta t}\right) \tag{6.2.14}$$

where

$$\beta = \epsilon\left(\frac{9A^4}{16}\right). \tag{6.2.15}$$

Integrating Eq. (6.2.14) and then imposing the condition $\phi(0,\epsilon) = 0$, produces the result

$$\phi(t,\epsilon) = -\left(\frac{\sqrt{3}A}{4}\right)t + \left(\frac{\sqrt{3}A}{4\beta}\right)\ln(1+\beta t), \tag{6.2.16}$$

and

$$\psi(t,\epsilon) = \Omega t + \phi(t,\epsilon) = \left(\frac{\sqrt{3}A}{4}\right)t + \left(\frac{\sqrt{3}A}{4\beta}\right)\ln(1+\beta t). \tag{6.2.17}$$

Therefore, an approximation to the solution of

$$\ddot{x} + x^3 = -\epsilon\dot{x}^3$$

is

$$x(t) \simeq \frac{A\cos\left[\left(\frac{\sqrt{3}A}{4}\right)t + \left(\frac{\sqrt{3}A}{4\beta}\right)\ln(1+\beta t)\right]}{\sqrt{1+\beta t}}. \tag{6.2.18}$$

6.2.3 $\ddot{x} + x^3 = \epsilon(1 - x^2)\dot{x}$

The above equation is the Duffing-van der Pol equation. For this case $F(x, \dot{x})$ is

$$F(x, \dot{x}) = (1 - x^2)\dot{x}, \tag{6.2.19}$$

and

$$F(x, \dot{x}) \to F(a\cos\psi, -\Omega a \sin\psi) = [1 - a^2(\cos\psi)^2](-\Omega a \sin\psi).$$

With this result, a direct and easy calculation gives the following expressions

$$\dot{a} = \left(\frac{\epsilon}{2}\right) a \left[1 - \frac{a^2}{4}\right], \tag{6.2.20}$$

$$\dot{\phi} = \left(\frac{1}{2\Omega}\right)\left[\frac{3a^2}{4} - \Omega^2\right]. \tag{6.2.21}$$

If $H_1(a)$ is taken to be the polynomial in the variable a, on the right-hand side of Eq. (6.2.20), then

$$H_1(A^*) = A^*\left[1 - \frac{A^{*2}}{4}\right] = 0 \Rightarrow A^* = 0 \text{ or } 2. \tag{6.2.22}$$

The first value, $A^* = 0$, is the equilibrium state, while the second value, $A^* = 2$, corresponds to the amplitude of a limit-cycle.

Equation (6.2.20) has the solution

$$a(t, \epsilon) = \frac{2A}{[A^2 + (4 - A^2)e^{-\epsilon t}]^{1/2}}, \tag{6.2.23}$$

and with this result Eq. (6.2.21) becomes

$$\dot{\phi} = \left(\frac{\sqrt{3}A}{4}\right)(4 - A^2)\left[\frac{1 - e^{-\epsilon t}}{A^2 + (4 - A^2)e^{-\epsilon t}}\right]. \tag{6.2.24}$$

Using the integral relations

$$\int \frac{dt}{c_1 + c_2 e^{-\epsilon t}} = \left(\frac{1}{\epsilon c_1}\right)\ln[c_2 + c_1 e^{\epsilon t}]$$

$$\int \frac{e^{-\epsilon t}dt}{c_1 + c_2 e^{-\epsilon t}} = -\left(\frac{1}{c_2 \epsilon}\right)\ln[c_1 + c_2 e^{-\epsilon t}],$$

and requiring $\phi(0, \epsilon) = 0$, gives the following expression for $\phi(t, \epsilon)$

$$\phi(t, \epsilon) = \left(\frac{\sqrt{3}A}{4}\right)\left\{\left(\frac{4 - A^2}{\epsilon A^2}\right)\ln[(4 - A^2) + A^2 e^{\epsilon t}]\right.$$

$$+\ln[A^2+(4-A^2)e^{-\epsilon t}]\Big\} - \left[\frac{\sqrt{3}\ln(4)}{4cA}\right][4-(1-\epsilon)A^2]. \quad (6.2.25)$$

Finally, the averaging approximation of Mickens and Oyedeji applied to the Duffing-van der Pol equation gives the following expression for the oscillatory solutions

$$x(t,\epsilon) \simeq a(t,\epsilon)\cos[\psi(t,\epsilon)],$$

$$\psi(t,\epsilon) = \left(\frac{\sqrt{3}A}{4}\right)t + \phi(t,\epsilon),$$

where $a(t,\epsilon)$ and $\phi(t,\epsilon)$ are the functions given, respectively, in Eqs. (6.2.23) and (6.2.25).

6.2.4 $\ddot{x} + x^{1/3} = -2\epsilon\dot{x}$

The linearly damped, cube-root TNL oscillator differential equation can be linearized by using the following expansion for $(\cos\theta)^{1/3}$ [12]

$$(\cos\theta)^{1/3} = a_1\cos\theta + a_2\cos 3\theta + a_3\cos 5\theta + \cdots.$$

Therefore

$$\begin{aligned} x^{1/3} \to (a\cos\theta)^{1/3} &= a^{1/3}(\cos\theta)^{1/3} = a^{1/3}[a_1\cos\theta + \text{HOH}] \\ &= \left[\frac{a_1}{a^{2/3}}\right](a\cos\theta) + \text{HOH}, \end{aligned}$$

and the harmonic linearization of this term is

$$x^{1/3} \to \left[\frac{a_1}{a^{2/3}}\right]x = \Omega^2(a)x, \quad (6.2.26)$$

with $a_1 = 1.15960$. This implies that a solution must be found for the equation

$$\ddot{x} + \Omega^2(a)x = -2\epsilon\dot{x}, \quad (6.2.27)$$

where the initial conditions are taken as

$$x(0) = A, \quad \dot{x}(0) = 0. \quad (6.2.28)$$

If in $\Omega^2(a)$, the a is replaced by A, then the resulting equation is a linear, second-order differential equation with constant coefficients. The exact solution to the equation

$$\ddot{y} + \Omega^2(A)y = -2\epsilon\dot{y}, \quad (6.2.29)$$

is

$$y(t,\epsilon) = c_1 e^{-\epsilon t}\cos\left\{\left[\Omega^2(A) - \epsilon^2\right]^{1/2} t + \phi\right\}$$

where c_1 and ϕ are integration constants. Therefore, to terms of order ϵ,

$$c_1 = A, \quad \phi = 0,$$

and

$$y(t,\epsilon) = Ae^{-\epsilon t}\cos\left\{[\Omega(A)t]t\right\}, \tag{6.2.30}$$

$$\Omega^2(A) = \frac{a_1}{A^{2/3}}, \tag{6.2.31}$$

and the approximation to the oscillatory solution of the linearly damped, cube-root equation is

$$x(t,\epsilon) \simeq y(t,\epsilon).$$

6.2.5 $\ddot{x} + x^{1/3} = \epsilon(1 - x^2)\dot{x}$

This equation is the cube-root/van der Pol differential equation and the reformulated equation to be studied is

$$\ddot{x} + \Omega^2 x = \epsilon(1 - x^2)\dot{x} \tag{6.2.32}$$

where, for the moment, we do not indicate the particular amplitude value upon which Ω^2 depends. Assuming Ω^2 is a constant, then a first-order averaging method gives

$$\frac{da}{dt} = \left(\frac{\epsilon}{2}\right) a \left[1 - \frac{a^2}{4}\right], \tag{6.2.33}$$

$$\frac{d\phi}{dt} = 0, \tag{6.2.34}$$

where

$$x(t,\epsilon) = a(t,\epsilon)\cos[\Omega t + \phi(t,\epsilon)]. \tag{6.2.35}$$

The right-side of Eq. (6.2.33) is zero for $a = 0$ and $a = 2$. As in Section 6.2.3, $a = 0$ is the equilibrium state and $a = 2$ corresponds to a limit-cycle. These results suggest that the angular frequency be evaluated at an amplitude equal to two, i.e.,

$$\Omega^2(2) = \frac{a_1}{2^{2/3}} = 0.7305. \tag{6.2.36}$$

Since the solution to Eq. (6.2.33) is known, see Eq. (6.2.23), and since $\phi(t,\epsilon)=0$, the combined linearization-averaging method gives the following approximate solution

$$x(t,\epsilon) \simeq \frac{2A\cos[(0.8547)t]}{[A^2+(4-A^2)e^{-\epsilon t}]^{1/2}}. \tag{6.2.37}$$

Let us now compare a harmonic balance calculation to the result given by Eq. (6.2.37). For

$$x = A\cos\theta, \quad \theta = \Omega t,$$

substituted in

$$\ddot{x} + x^{1/3} = \epsilon(1-x^2)\dot{x},$$

we obtain the result

$$(-\Omega^2 A + A^{1/3}a_1)\cos\theta + (\epsilon\Omega A)\left[1-\frac{A^2}{4}\right]\sin\theta + \text{HOH} \simeq 0.$$

If the coefficients of $\cos\theta$ and $\sin\theta$ are set to zero, we obtain

$$\Omega^2 = \frac{a_1}{A^{2/3}}; \quad A = 0 \text{ or } 2.$$

For $A=2$, it follows that

$$\Omega^2 = \frac{a_1}{2^{2/3}} = 0.7305$$

and

$$x(t) = 2\cos[(0.8547)t]. \tag{6.2.38}$$

This is exactly the result obtained if t is taken to be large in Eq. (6.2.37).

6.2.6 $\ddot{x} + x = -2\epsilon(\dot{x})^{1/3}$

This differential equation is the "fractional damped" linear harmonic oscillator [13]. It is of interest to investigate the solutions to this equation although it is not strictly speaking a TNL oscillator.

The standard first-order averaging method [3] can be applied to

$$\ddot{x} + x = -2\epsilon(\dot{x})^{1/3}, \tag{6.2.39}$$

and the following equations are obtained

$$\frac{da}{dt} = -\left(\frac{\epsilon}{2\pi}\right)a^{1/3}\int_0^{2\pi}(\sin\psi)^{4/3}d\psi = -(\epsilon c_0)a^{1/3}, \tag{6.2.40}$$

$$\frac{d\phi}{dt} = -\left(\frac{\epsilon}{2\pi a}\right) a^{1/3} \int_0^{2\pi} (\sin\psi)^{1/3} \cos\psi \, d\psi = 0. \tag{6.2.41}$$

For this problem

$$F(x, \dot{x}) = -2(\dot{x})^{1/3} \to (-2)(-a\sin\psi)^{1/3},$$

and [19]

$$(\sin\psi)^{1/3} = c_0 \sin\psi + c_1 \sin 3\psi + \cdots$$

where $c_0 = 0.579796$. With the initial conditions

$$a(0, \epsilon) = A, \quad \phi(0, \epsilon) = 0,$$

the solutions of Eqs. (6.2.40) and (6.2.41) are

$$a(t, \epsilon) = \begin{cases} A\left(\dfrac{t^* - t}{t^*}\right)^{3/2}, & 0 \le t \le t^* \\ 0, & t > t^*, \end{cases} \tag{6.2.42}$$

where

$$t^* = \frac{3A^{2/3}}{2c_0\epsilon}. \tag{6.2.43}$$

Therefore, an approximation to the oscillatory solution of Eq. (6.2.39) is the following expression

$$x(t, \epsilon) = \begin{cases} A\left(\dfrac{t^* - t}{t^*}\right)^{3/2} \cos t, & 0 \le t \le t^*, \\ 0, & t > t^*. \end{cases} \tag{6.2.44}$$

The calculations presented above demonstrate that the "fractional damped," linear harmonic oscillator undergoes only a finite number of oscillations, and these take place in a time interval equal to t^*. If $N(A, \epsilon)$ is the number of these oscillations, then [7, 13]

$$2\pi N \simeq t^*$$

and

$$N(A, \epsilon) \simeq \frac{3A^{2/3}}{4\pi c_0 \epsilon}. \tag{6.2.45}$$

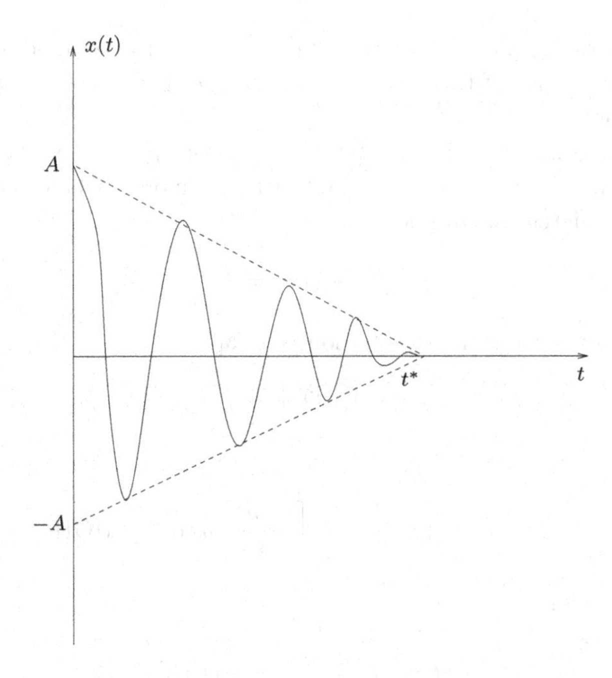

Fig. 6.2.1 Schematic representation of the solution for $\ddot{x} + x = -2\epsilon(\dot{x})^{1/3}$.

6.2.7 *General Comments*

- Clearly, both the Mickens-Oyedeji [4] and Mickens [5] procedures provide accurate representations with respect to the overall qualitative features of the solutions to nonconservative TNL oscillator differential equations.
- For the TNL oscillator

$$\ddot{x} + x^3 = \epsilon F(x, \dot{x}), \quad 0 < \epsilon \ll 1,$$

a better choice for selecting Ω^2 is to replace it, not with $\Omega^2 = 3A^2/4$, but with the exact value (for $\epsilon = 0$)

$$\Omega^2_{\text{exact}}(A) = \left[\frac{\pi A}{2F\left(\frac{1}{\sqrt{2}}\right)}\right]^2 = (0.7177705)A^2,$$

where $F(k)$ is the complete elliptic function of the first kind [12].
- In the worked examples, for which the Mickens-Oyedeji procedure was applied, we did not do the calculations as presented in [4]. The current method provides a mechanism for incorporating time dependency

into the $\phi(t,\epsilon)$ part of the phase function $\psi(t,\epsilon) = \Omega t + \phi(t,\epsilon)$. A major consequence of this change is that $\psi(t,\epsilon)$ now has a complex time behavior.

- The harmonic linearization for the function $g(x)$ is exactly what is determined by carrying out a first-order harmonic balance calculation for the differential equation

$$\ddot{x} + g(x) = 0,$$

i.e., the equivalent linear equation is [1–3]

$$\ddot{x} + \Omega^2 x = 0,$$

where

$$g(x) \to g(a\cos\theta) = \left[\frac{g_0(a)}{a}\right] a\cos\theta + \text{HOH}$$

and, therefore

$$g(x) \to \left[\frac{g_0(a)}{a}\right] x = \Omega^2(a)x.$$

- Both the Mickens-Oyedeji [4] and Mickens [5] procedures allow the calculation of approximate values for the amplitude and angular frequency of existing limit-cycles, as well as the transitory behavior in the approach to these periodic solutions.

6.3 Cveticanin's Averaging Method

A general and powerful extension of the Mickens-Oyedeji [4] and Mickens [5] methods was constructed by Cveticanin [6]. This procedure starts with the calculation of the exact angular frequency for the equation

$$\ddot{x} + |x|^\alpha \text{sgn}(x) = 0, \tag{6.3.1}$$

and then uses this result to derive a first-order (in ϵ) averaging method for

$$\ddot{x} + |x|^\alpha \text{sgn}(x) = \epsilon F(x, \dot{x}). \tag{6.3.2}$$

This section presents the details of these calculations. Our presentation follows closely the work as given in Cveticanin's publication [6].

6.3.1 *Exact Period*

Consider the TNL oscillator equation

$$\ddot{x} + c_1^2 x|x|^{\alpha-1} = 0, \tag{6.3.3}$$

with the initial conditions

$$x(0) = A, \quad \dot{x}(0) = 0, \tag{6.3.4}$$

and c_1 a real constant. Note that Eq. (6.3.3) is an alternative form of Eq. (6.3.1).

The system equations for Eq. (6.3.3) are

$$\frac{dx}{dt} = y, \quad \frac{dy}{dt} = -c_1^2 x|x|^{\alpha-1}, \tag{6.3.5}$$

and, by inspection, it follows that $(\bar{x}, \bar{y}) = (0,0)$ is the location of the fixed-point in the two-dim (x, y) phase space.

The first-order differential equation for the trajectories, $y = y(x)$, in phase space is

$$\frac{dy}{dx} = -\frac{c_1^2 x|x|^{\alpha-1}}{y}, \tag{6.3.6}$$

and this separable equation can be solved to give a first-integral for Eq. (6.3.3), i.e.,

$$\frac{y^2}{2} + \left(\frac{c_1^2}{\alpha+1}\right)|x|^{\alpha+1} = \left(\frac{c_1^2}{\alpha+1}\right)A^{\alpha+1}, \tag{6.3.7}$$

where we take $A \geq 0$. Since both terms on the left-hand side are non-negative, we conclude that Eq. (6.3.7) corresponds to a simple, closed curve in the (x, y) phase-plane. Consequently, all solutions to Eq. (6.3.3) are periodic [19].

The phase-space trajectory that starts at $(A, 0)$ and lies in the fourth quadrant corresponds to one-fourth of the full, closed trajectory. Using Eq. (6.3.7), it follows that the period is given by the expression [6, 15–18]

$$\frac{T(A)}{4} = \left(\frac{\alpha+1}{2c_1^2}\right)^{1/2} \int_0^A \frac{dx}{\sqrt{A^{\alpha+1} - x^{\alpha+1}}}. \tag{6.3.8}$$

The following change of variable

$$x = Au^{\frac{1}{\alpha+1}}, \tag{6.3.9}$$

transforms Eq. (6.3.8) into the form

$$T(A) = \left(\frac{4A^{\frac{1-\alpha}{2}}}{c_1\sqrt{2(\alpha+1)}}\right) \int_0^1 (1-u)^{-1/2} u^{-\left(\frac{\alpha}{\alpha+1}\right)} du. \tag{6.3.10}$$

Using the definition of the beta function [19]

$$B(p,q) \equiv \int_0^1 (1-u)^{p-1} u^{q-1} du = \frac{\Gamma(p)\Gamma(q)}{\Gamma(p+q)}, \tag{6.3.11}$$

where $\Gamma(z)$ is the gamma function [19], we find that

$$p = \frac{1}{2}, \qquad q = \frac{1}{\alpha+1}. \tag{6.3.12}$$

Therefore,

$$T(A) = \left(\frac{4A^{\frac{1-\alpha}{2}}}{c_1\sqrt{2(\alpha+1)}} \right) \frac{\Gamma\left(\frac{1}{2}\right)\Gamma\left(\frac{1}{\alpha+1}\right)}{\Gamma\left[\frac{\alpha+3}{2(\alpha+1)}\right]}, \tag{6.3.13}$$

and the angular frequency is

$$\Omega(A) = \frac{2\pi}{T(A)} = c_1 \left[\frac{\pi(\alpha+1)}{2}\right]^{1/2} \left\{ \frac{\Gamma\left[\frac{\alpha+3}{2(\alpha+1)}\right]}{\Gamma\left(\frac{1}{\alpha+1}\right)} \right\} A^{\frac{\alpha-1}{2}}. \tag{6.3.14}$$

To obtain this result, we used $\Gamma\left(\frac{1}{2}\right) = \sqrt{\pi}$. Note that $\Omega(A)$ is only defined for $\alpha > -1$.

6.3.2 *Averaging Method* [6]

The averaging method of Cveticanin [6], for the TNL oscillator

$$\ddot{x} + c_1^2 x|x|^{\alpha-1} = \epsilon F(x, \dot{x}), \quad 0 < \epsilon \ll 1, \tag{6.3.15}$$

begins with the assumption that the exact solution can be written as

$$x(t,\epsilon) = a(t,\epsilon)\cos\psi(t,\epsilon) \tag{6.3.16}$$

where

$$\psi(t,\epsilon) = \int \Omega(a)dt + \beta(t), \tag{6.3.17}$$

with $\Omega(a)$ taken from Eq. (6.3.14). With the condition (see Appendix G)

$$\dot{a}\cos\psi - a\dot{\beta}\sin\psi = 0, \tag{6.3.18}$$

the first-derivative, $\dot{x}$, is

$$\dot{x} = -a\Omega(a)\sin\psi. \tag{6.3.19}$$

Substituting Eqs. (6.3.16) and (6.3.19) into Eq. (6.2.15) and carrying out the indicated mathematical operations gives

$$\begin{aligned} &\dot{a}\Omega\sin\psi + a\dot{a}\Omega'\sin\psi \\ &\quad + a\Omega\dot{\beta}\cos\psi = -\epsilon F(a\cos\psi, -a\Omega\sin\psi), \end{aligned} \tag{6.3.20}$$

where

$$\Omega'(a) \equiv \frac{d\Omega(a)}{da}, \tag{6.3.21a}$$

and

$$\dot{\Omega}(a) = \Omega'(a)\dot{a}. \tag{6.3.21b}$$

Inspection of Eqs. (6.3.18) and (6.3.20) shows that they are linear in both $\dot{a}$ and $a\dot{\beta}$. Therefore solving for these quantities provides the following relations

$$\left[1 + \left(\frac{\alpha - 1}{2}\right)(\sin\psi)^2\right]\dot{a} = -\left(\frac{\epsilon}{\Omega}\right)F\sin\psi,$$

$$a\dot{\beta} + \dot{a}\left(\frac{\alpha - 1}{4}\right)\sin 2\psi = -\left(\frac{\epsilon}{\Omega}\right)F\cos\psi,$$

where $F = F(a\cos\psi, -a\Omega\sin\psi)$. These two expressions may be rewritten to the form

$$\left(\frac{\alpha + 3}{4}\right)\dot{a} = \left(\frac{\alpha - 1}{4}\right)\dot{a}\cos 2\psi - \left(\frac{\epsilon}{\Omega}\right)F\sin\psi, \tag{6.3.22}$$

$$a\dot{\beta} = -\left(\frac{\alpha - 1}{4}\right)\dot{a}\sin 2\psi - \left(\frac{\epsilon}{\Omega}\right)F\cos\psi. \tag{6.3.23}$$

Up to now, no approximations have been made, i.e., Eqs. (6.3.22) and (6.3.23) are the exact differential equations for $a(t,\epsilon)$ and $\beta(t,\epsilon)$. However, in general, there is little chance that they can be solved exactly. We proceed by observing that the right-sides of Eqs. (6.3.22) and (6.3.23) are periodic in ψ with period 2π. Therefore, averaging over 2π, i.e.,

$$\text{average} \equiv \left(\frac{1}{2\pi}\right)\int_0^{2\pi}(\cdots)d\psi,$$

gives for $\dot{a}$ and $\dot{\beta}$, the relations

$$\dot{a} = -\left[\frac{2\epsilon}{\pi(\alpha+3)\Omega(a)}\right]\int_0^{2\pi} F(a\cos\psi, -a\Omega(a)\sin\psi)\sin\psi\, d\psi, \tag{6.3.24}$$

$$\dot{\psi} = \Omega(a) - \left[\frac{\epsilon}{2\pi a\Omega(a)}\right]\int_0^{2\pi} F(a\cos\psi, -a\Omega(a)\sin\psi)\cos\psi\, d\psi, \tag{6.3.25}$$

and they are to be solved for the following initial conditions

$$a(0,\epsilon) = A, \quad \psi(0,\epsilon) = 0. \tag{6.3.26}$$

6.3.3 *Summary*

The averaging method proposed by Cveticanin [6] is a generalization of the Mickens-Oyedeji procedure [4] and is a major improvement over the Mickens combined linearization-averaging technique. It permits the direct calculation of an approximation to the oscillatory solutions of

$$\ddot{x} + c_1^2 x|x|^{\alpha-1} = \epsilon F(x, \dot{x}), \quad 0 < \epsilon \ll 1,$$

by means of Eqs. (6.3.24) and (6.3.25). The basis of the method is having an exact formula for the period or angular frequency when $\epsilon = 0$.

Interestingly, Cveticanin does not calculate solutions for situations where limit-cycles may exist. In the next section, we preform such determinations.

6.4 Worked Examples

6.4.1 $\ddot{x} + x|x|^{\alpha-1} = -2\epsilon\dot{x}$

The above TNL differential equation is a linearly damped, conservative TNL oscillator. Its averaged equations for the amplitude, $a(t,\epsilon)$, and phase, $\psi(t,\epsilon)$, are

$$\begin{aligned}\dot{a} &= -\left[\frac{2\epsilon}{\pi(\alpha+3)\Omega(A)}\right]\int_0^{2\pi}[(-)(-2a\Omega\sin\psi)]\sin\psi\, d\psi \\ &= -\left[\frac{4\epsilon a}{\pi(\alpha+3)}\right]\int_0^{2\pi}(\sin\psi)^2 d\psi = -\left(\frac{4\epsilon}{\alpha+3}\right)a, \end{aligned} \tag{6.4.1}$$

$$\begin{aligned}\dot{\psi} &= \Omega(a) - \left[\frac{\epsilon}{2\pi a\Omega(a)}\right]\int_0^{2\pi}(-)(-2a\Omega\sin\psi)\cos\psi\, d\psi \\ &= \Omega(a) - \left(\frac{\epsilon a}{2\pi}\right)\int_0^{2\pi}\sin\psi\cos\psi\, d\psi = \Omega(a), \end{aligned} \tag{6.4.2}$$

where

$$F(x,\dot{x}) = -2\dot{x} \to (-2)(-a\Omega\sin\psi). \tag{6.4.3}$$

The solution to Eq. (6.4.1), for $a(0,\epsilon) = A$, is

$$a(t,\epsilon) = Ae^{-\left(\frac{4\epsilon}{\alpha+3}\right)t}. \tag{6.4.4}$$

Substituting this expression for $a(t,\epsilon)$ into Eq. (6.4.2) gives

$$\dot{\psi} = \Omega(a) = qa^{\frac{\alpha-1}{2}}, \tag{6.4.5}$$

where

$$q = c_1 \left[\frac{(\alpha+1)\pi}{2}\right]^{1/2} \frac{\Gamma\left[\frac{3+\alpha}{2(\alpha+1)}\right]}{\Gamma\left(\frac{1}{\alpha+1}\right)}, \tag{6.4.6}$$

or

$$\dot{\psi} = \left[qA^{\frac{\alpha-1}{2}}\right] \exp\left\{-\left[\frac{2\epsilon(\alpha-1)t}{\alpha+3}\right]\right\}. \tag{6.4.7}$$

The solution to Eq. (6.4.7), subject to $\psi(0,\epsilon) = 0$, is

$$\psi(t,\epsilon) = -\left[\frac{\alpha+3}{2\epsilon(\alpha-1)}\right] qA^{\frac{\alpha-1}{2}} \left\{\exp\left(-\left[\frac{2\pi(\alpha-1)t}{(\alpha+3)}\right]\right) - 1\right\}. \tag{6.4.8}$$

Therefore, it follows that a first averaging approximation for the solution to the linearly damped TNL oscillator equation is

$$\begin{aligned} x(t,\epsilon) = A&\left\{\exp\left[-\left(\frac{4\epsilon t}{\alpha+3}\right)\right]\right\} \\ &\cdot \cos\left[\frac{\alpha+3}{2\epsilon(\alpha-1)}\right]\left[qA^{\frac{\alpha-1}{2}}\right]\left\{1 - \exp\left(-\left[\frac{2\epsilon(\alpha-1)t}{(\alpha+3)}\right]\right)\right\}. \end{aligned} \tag{6.4.9}$$

To gauge the accuracy of the averaging method in ϵ, we examine the case for $\alpha = 1$, i.e.,

$$\ddot{x} + c_1^2 x = -2\epsilon\dot{x}. \tag{6.4.10}$$

The exact solution is

$$x(t,\epsilon) = De^{-\epsilon t}\cos\left[\left(\sqrt{c_1^2 - \epsilon^2}\right)t\right], \tag{6.4.11}$$

while the averaging approximation solution is, from Eq. (6.4.10), the expression

$$x(t,\epsilon) = Ae^{-\epsilon t}\cos(c_1 t). \tag{6.4.12}$$

Comparison of these two solutions shows us that the averaged derived solution is correct to terms of $O(\epsilon)$. This result is consistent with the nature of the averaging procedure [1–3, 6].

6.4.2 $\ddot{x} + x|x|^{\alpha-1} = -2\epsilon(\dot{x})^3$

This case corresponds to a nonlinear, cubic damped, TNL oscillator. We have

$$\begin{aligned} F(x, \dot{x}) &= -2(\dot{x})^3 \to (-2)(-a\Omega \sin\psi)^3 \\ &= \left(\frac{a^3\Omega^3}{2}\right)(3\sin\psi - \sin 3\psi), \end{aligned} \tag{6.4.13}$$

and

$$\begin{aligned} \dot{a} &= -\left[\frac{2\epsilon}{\pi(\alpha+3)\Omega}\right]\int_0^{2\pi}\left(\frac{a^3\Omega^3}{2}\right)(3\sin\psi - \sin 3\psi)\sin\psi\, d\psi \\ &= -\left[\frac{\epsilon a^3\Omega^2}{\pi(\alpha+3)}\right]\left(\frac{3}{2}\right)(2\pi) = -\left(\frac{3\epsilon a^3\Omega^2}{(\alpha+3)}\right). \end{aligned} \tag{6.4.14}$$

Using

$$\Omega(a) = qa^{\frac{\alpha-1}{2}},$$

we obtain

$$a^3\Omega^2 = q^2 a^{\alpha+2},$$

and

$$\dot{a} = -\left[\frac{3\epsilon q^2}{(\alpha+3)}\right] a^{\alpha+2}. \tag{6.4.15}$$

The solution to the last equation, subject to the condition $a(0, \epsilon) = A$, is

$$a(t, \epsilon) = \frac{A}{\left[1 + 3\epsilon\left(\frac{\alpha+1}{\alpha+3}\right) q^2 A^{(\alpha+1)} t\right]^{\frac{1}{\alpha+1}}}. \tag{6.4.16}$$

Similarly, for $\dot{\psi}$, we find

$$\dot{\psi} = \Omega(a) = qa^{\frac{\alpha-1}{2}}. \tag{6.4.17}$$

Since

$$\int \frac{dt}{(1 + D_2 t)^{D_3}} = \left[\frac{1}{D_2(1 - D_3)}\right]\frac{1}{(1 + D_2 t)^{(D_3 - 1)}}, \tag{6.4.18}$$

it follows that

$$\begin{aligned} \psi(t) = &-\left(\frac{2}{3\epsilon q}\right)\left[\frac{1}{A^{\left(\frac{\alpha+3}{2}\right)}}\right] \\ &\cdot \left\{1 - \left[1 + 3\epsilon q^2 A^{(\alpha+1)}\left(\frac{\alpha+1}{\alpha+3}\right)t\right]^{\frac{(\alpha+3)}{2(\alpha+1)}}\right\}. \end{aligned} \tag{6.4.19}$$

Finally, the first-order averaged solution to

$$\ddot{x} + x|x|^{\alpha-1} = -2\epsilon(\dot{x})^3, \tag{6.4.20}$$

is

$$x(t,\epsilon) = a(t,\epsilon)\cos\psi(\epsilon,t), \tag{6.4.21}$$

with $a(t,\epsilon)$ and $\psi(\epsilon,t)$ given, respectively by Eqs. (6.4.16) and (6.4.19).

Note that for $\alpha = 3$, we have

$$a(t,\epsilon) = \frac{A}{[1+(2\epsilon q^2 A^4)t]^{1/4}}, \tag{6.4.22}$$

and

$$\psi(t,\epsilon) = -\left[\frac{2}{3\epsilon q A^3}\right]\left\{1 - [1+(2\epsilon q^2 A^4)t]^{3/4}]\right\}. \tag{6.4.23}$$

These results are to be compared to the finding in Section 6.2.2; i.e.,

$$a(t,\epsilon) = \frac{A}{\left[1+\epsilon\left(\frac{9A^4}{16}\right)t\right]^{1/2}}, \tag{6.4.24}$$

and

$$\psi(t,\epsilon) = \left(\frac{\sqrt{3}A}{A}\right)t + \left(\frac{\sqrt{3}A}{4\beta}\right)\ln(1+\beta t), \tag{6.4.25}$$

where

$$\beta = \epsilon\left(\frac{9A^4}{16}\right). \tag{6.4.26}$$

Obviously, major differences exist in the predictions of the Mickens-Oyedeji methods [4] and the current technique [6].

6.4.3 $\ddot{x} + x|x|^{\alpha-1} = \epsilon(1-x^2)\dot{x}$

This equation is a modified version of the standard van der Pol oscillator differential equation

$$\ddot{x} + x = \epsilon(1-x^2)\dot{x}. \tag{6.4.27}$$

The first-order averaged solution for Eq. (6.4.27) is [3]

$$x(\epsilon,t) = \frac{2A\cos t}{[A^2+(4-A^2)e^{-\epsilon t}]}^{1/2}. \tag{6.4.28}$$

For the TNL oscillator

$$\ddot{x} + x|x|^{\alpha-1} = \epsilon(1 - x^2)\dot{x}, \tag{6.4.29}$$

we have

$$F(x, \dot{x}) = (1 - x^2)\dot{x}, \tag{6.4.30}$$

and

$$\begin{aligned} F(a\cos\psi, -a\Omega\sin\psi)\sin\psi &= [1 - a^2(\cos\psi)^2](-a\Omega\sin\psi)\sin\psi \\ &= -\left(\frac{a\Omega}{2}\right)\left[1 - \frac{a^2}{4}\right] + \text{HOH}. \end{aligned} \tag{6.4.31}$$

Therefore, from Eq. (6.3.24), we have

$$\dot{a} = \left(\frac{2\epsilon}{\alpha+3}\right) a \left[1 - \frac{a^2}{4}\right]. \tag{6.4.32}$$

A direct calculation for $\dot{\psi}$, see Eq. (6.3.25) gives

$$\dot{\psi} = \Omega(a) = q a^{\left(\frac{\alpha-1}{2}\right)}. \tag{6.4.33}$$

The solution to Eq. (6.4.32) is

$$a(t, \epsilon) = \frac{2A}{\left\{A^2 + (4 - A^2)\exp\left[-\left(\frac{4\epsilon}{\alpha+3}\right)t\right]\right\}^{1/2}}, \tag{6.4.34}$$

and substituting this function for $a(t, \epsilon)$ into Eq. (6.4.33) gives

$$\dot{\psi} = \left[q(2A)^{\frac{\alpha-1}{2}}\right] \frac{1}{\left\{A^2 + (4 - A^2)\exp\left[-\left(\frac{4\epsilon}{\alpha+3}\right)t\right]\right\}^{\left(\frac{\alpha-1}{4}\right)}}. \tag{6.4.35}$$

In general, this equation cannot be integrated in closed form for arbitrary values of α. However, for $\alpha = 1$, i.e., the standard van der Pol equation with $c_1 = 1$, the result given by Eq. (6.4.28) is found. Also, one consequence of Eq. (6.4.34) is

$$\lim_{t\to\infty} a(t, \epsilon) = 2. \tag{6.4.36}$$

This implies that for large t, the oscillatory solutions to Eq. (6.4.29) approach a stable limit-cycle having amplitude $a(\infty, \epsilon) = 2$. Further, for large t, it follows from Eq. (6.4.33) that the phase is

$$\psi(t, \epsilon) = \left[q 2^{\left(\frac{\alpha-1}{2}\right)}\right] t. \tag{6.4.37}$$

For $\alpha = 3$ and $c_1 = 1$, i.e.,

$$\ddot{x} + x^3 = \epsilon(1 - x^2)\dot{x}, \tag{6.4.38}$$

we have

$$x(t, \epsilon) \underset{t \text{ large}}{\longrightarrow} 2\cos(1.6954426\, t) \tag{6.4.39}$$

where

$$q = \frac{\sqrt{2\pi}\,\Gamma\left(\frac{3}{4}\right)}{\Gamma\left(\frac{1}{4}\right)} = (0.8477213).$$

This result is to be compared to the value derived from first-order harmonic balance

$$x_{HB}(t, \epsilon) \underset{t \text{ large}}{\longrightarrow} 2\cos(\sqrt{3}\, t) = 2\cos(1.7320508t). \tag{6.4.40}$$

6.5 Chronology of Averaging Methods

The 1943 book by Krylov and Bogoliubov [1] was the first public description of the first-order averaging method. Following the 1985 article by Mickens and Oyedeji [4] on the construction of an averaging procedure for the TNL oscillator equation

$$\ddot{x} + x^3 = \epsilon F(x, \dot{x}), \tag{6.5.1}$$

a number of researchers created a broad range of related, but generalized techniques for investigating the non-steady state solutions of oscillator equations taking the form

$$\ddot{x} + g(x) = \epsilon F(x, \dot{x}), \tag{6.5.2}$$

where $g(x)$ can be

$$g(x) \to \begin{cases} x^3, \\ ax + bx^3, \\ x|x|, \\ \text{etc.,} \end{cases} \tag{6.5.3}$$

and possible functions for $F(x, \dot{x})$ include

$$F(x, \dot{x}) \to \begin{cases} -\dot{x}, \\ -(\dot{x})^3, \\ (1 - x^2)\dot{x}, \\ \text{etc.} \end{cases} \tag{6.5.4}$$

The following is a partial chronological listing of some of the significant publications on this topic:

1. N. Krylov and N. Bogoliubov, *Introduction to Nonlinear Mechanics* (Princeton University Press; Princeton, NJ; 1943).
2. R. E. Mickens and K. Oyedeji, "Construction of approximate analytical solutions to a new class of nonlinear oscillator equation," *Journal of Sound and Vibration* **102**, 579–582 (1985).
3. S. B. Yuste and J. D. Bejarano, "Construction of approximation analytical solutions to a new class of nonlinear oscillator equation," *Journal of Sound and Vibration* **110**, 347–350 (1986).
4. S. B. Yuste and J. D. Bejarano, "Improvement of a Krylov-Bogoliubov method that uses Jacobi elliptic function," *Journal of Sound and Vibration* **139**, 151–163 (1990).
5. V. T. Coppola and R. H. Rand, "Averaging using elliptic functions: Approximation of limit cycles," *Acta Mechanica* **81**, 125–142.
6. Z. Xu and Y. K. Cheung, "Averaging method using generalized harmonic functions for strongly nonlinear oscillators," *Journal of Sound and Vibration* **174**, 563–576 (1994).
7. S. H. Chen, X. M. Yang, and Y. K. Cheung, "Periodic solutions of strongly quadratic nonlinear oscillators by the elliptic perturbation method," *Journal of Sound and Vibration* **212**, 771–780 (1998).
8. L. Cveticanin, "Analytical methods for solving strongly nonlinear differential equations," *Journal of Sound and Vibration* **214**, 325–338 (1998).
9. A. Chatterjee, "Harmonic balance based averaging: Approximate realizations of an asymptotic techniques," *Nonlinear Dynamics* **32**, 323–343 (2003).
10. L. Cveticanin, "Oscillator with fraction order restoring," *Journal of Sound and Vibration* **320**, 1064 (2008).

The majority of the above studies focus on cubic nonlinearities, i.e.,

$$\ddot{x} + ax + bx^3 = \epsilon F(x, \dot{x}), \quad 0 < \epsilon \ll 1, \tag{6.5.5}$$

where (a, b) are constants. Since the differential equation

$$\ddot{x} + ax + bx^3 = 0, \tag{6.5.6}$$

can be exactly solved in terms of Jacobi elliptic function [20, 21], several of the averaging formulations are based on procedures which perturb off these functions.

Finally, the large investment in efforts to study cubic systems arises because such functions readily occur in the mathematical modeling of a broad range of systems appearing in the natural and engineering sciences [22–24].

6.6 Comments

- Truly nonlinear oscillator equations

$$\ddot{x} + g(x) = \epsilon F(x, \dot{x}), \quad 0 < \epsilon \ll 1, \tag{6.6.1}$$

have solutions where amplitudes and angular frequencies depend on time. In addition to purely dissipative systems, for which the amplitude decreases monotonously to zero, other dissipative systems may possess limit-cycles. Clearly, harmonic balance, parameter expansion, and iteration procedures cannot be used to capture the transient behavior of the solutions. This fact demonstrates the importance of having averaging methods for calculating these oscillator behaviors.

- The Krylov-Bogoliubov method was derived for application to standard nonlinear oscillator differential equations having the form [1–3]

$$\ddot{x} + x = \epsilon F(x, \dot{x}), \quad 0 < \epsilon \ll 1. \tag{6.6.2}$$

However, in its original form, this method cannot be directly applied to TNL oscillator differential equations, such as that given in Eq. (6.6.1). The first major generalization of the Krylov-Bogoliubov method was done by Mickens and Oyedeji [4]. However, their work focused only on the class of TNL equations for which $g(x)$ was a pure cubic expression, i.e., $g(x) = x^3$ and

$$\ddot{x} + x^3 = \epsilon F(x, \dot{x}), \quad 0 < \epsilon \ll 1. \tag{6.6.3}$$

For this procedure, a major issue is the selection of the angular frequency, Ω [4]. One possibility is to use a harmonic balance approximation for Ω. A second possibility is to replace Ω by its exact value (when this is known) for the conservative portion of the equation, i.e., the resulting differential equation for $\epsilon = 0$. In any case, the structure of the assumed solution is

$$\begin{cases} x(t,\epsilon) = a(t,\epsilon)\cos\psi(t,\epsilon), \\ \psi(t,\epsilon) = \Omega t + \phi(t,\epsilon), \end{cases} \tag{6.6.4}$$

where the approximations to the amplitude and phase functions satisfy the following two first-order differential equations (see Eqs. (6.1.10) and (6.1.11))

$$\dot{a} = -\left(\frac{\epsilon}{2\pi\Omega}\right)\int_0^{2\pi} F(a\cos\psi, -\Omega a\sin\psi)\sin\psi\, d\psi, \tag{6.6.5}$$

$$\dot{\phi} = \left(\frac{1}{2}\right)\left(\frac{3a^2}{4\Omega} - \Omega\right)$$

$$- \left(\frac{\epsilon}{2\pi\Omega a}\right) \int_0^{2\pi} F(a\cos\psi, -\Omega a \sin\psi)\cos\psi \, d\psi, \qquad (6.6.6)$$

where it should be understood that Ω is a function of a, i.e., $\Omega = \Omega(a)$. If $F(x, \dot{x})$ has the form

$$F(x, \dot{x}) = F_1(x^2, \dot{x}^2)\dot{x}, \qquad (6.6.7)$$

then the integral in Eq. (6.6.6) is zero and the relation involving $\dot{\phi}$ becomes

$$\dot{\phi} = \left(\frac{1}{2}\right)\left(\frac{3a^2}{4\Omega} - \Omega\right). \qquad (6.6.8)$$

The procedure for calculating $a(t, \epsilon)$ and $\phi(t, \epsilon)$ now becomes:

i) Carry out the integration in Eq. (6.6.5) to obtain

$$\dot{a} = \epsilon H(a). \qquad (6.6.9)$$

ii) Solve this first-order differential equation for $a = a(t, \epsilon)$, with $a(0, \epsilon) = A$.

iii) Substitute $a(t, \epsilon)$ into the right-hand side of Eq. (6.6.8) and solve for $\phi(t, \epsilon)$. With

$$\psi(t, \epsilon) = \Omega t + \phi(t, \epsilon),$$

and require $\psi(0, \epsilon) = 0$.

iv) Substitution of these functions into the first of Eq. (6.6.4) gives a first-order averaging solution to Eq. (6.6.3) according to the Mickens-Oyedeji procedure. (Note that the above implementation of the method differs from the original presentation [4].)

- The Mickens combined linearization-averaging method [4] is only expected to provide the general qualitative features of the solutions for TNL oscillators. Its use should be restricted to situations where only a quick, overall knowledge of the system's behavior is required.
- Finally, the Cveticanin methodology [6] is the proper generalization of the original Krylov-Bogoliubov first-order, averaging method [1, 2]. An advantage of this procedure is that it provides a clear and unambiguous set of rules for calculating approximations to the oscillatory solutions of TNL differential equations expressed in the form

$$\ddot{x} + c_1 x|x|^{\alpha-1} = \epsilon F(x, \dot{x}). \qquad (6.6.10)$$

Problems

6.1 Show that if

$$F(x, \dot{x}) = F_1(x, \dot{x}^2)\dot{x},$$

then

$$\int_0^{2\pi} F(a\cos\psi, -a\Omega\sin\psi)\cos\psi = 0.$$

6.2 Derive Eq. (6.1.7) and, Eqs. (6.1.8) and (6.1.9). Use these results to obtain Eqs. (6.1.10) and (6.1.11).

6.3 Apply the Bejarano-Yuste elliptic function perturbation method to the equations

$$\begin{aligned} \ddot{x} + x^3 &= -2\epsilon\dot{x} \\ \ddot{x} + x^3 &= -2\epsilon\dot{x}^3 \\ \ddot{x} + x^3 &= \epsilon(1 - x^2)\dot{x}. \end{aligned}$$

See [8].

6.4 Calculate harmonic linearizations for

$$g(x) = \begin{cases} x^{1/3} \\ x + x^{1/3} \\ x + x^3. \end{cases}$$

Plot the corresponding $\Omega^2(a)$ versus a.

6.5 Does $x^{-1/3}$ have a harmonic linearization? If so, calculate it. If not, provide an explanation for its non-existence.

6.6 Solve Eq. (6.2.6) to obtain the solution given in Eq. (6.2.7).

6.7 Calculate $\phi(t, \epsilon)$ for the differential equation given by Eq. (6.2.14).

6.8 Solve for $a(t, \epsilon)$ and $\phi(t, \epsilon)$ from Eqs. (6.2.20) and (6.2.21).

6.9 For

$$\ddot{x} + c_1 x|x|^{\alpha-1} = \epsilon(1 - x^2)\dot{x},$$

why are the amplitude differential equations the same for all allowable values of α for the Cveticanin [6] or Mickens [5] methods?

6.10 Consider the equation

$$\ddot{x} + x = -2\epsilon(\dot{x})^{1/3}.$$

Can a physical reason be provided to explain why the oscillatory behavior stops after only a finite number of oscillations?

6.11 Derive the Fourier expansion of $(\sin\psi)^{1/3}$. See Mickens [19].

6.12 What is the harmonic linearization of $(\dot{x})^{1/3}$?
6.13 Derive the result in Eqs. (6.2.42) and (6.2.43).
6.14 Explain in detail the reasoning that leads to Eq. (6.2.45).
6.15 Is the following relationship

$$x|x|^{\alpha-1} = |x|^{\alpha}\text{sgn},$$

always correct?
6.16 Derive Eq. (6.3.10) from Eq. (6.3.8), and show that $\Omega(A)$ has the value expressed in Eq. (6.3.14).
6.17 Show that the result for $\Omega(A)$, as presented in Eq. (6.3.14), holds only for $\alpha > -1$.
6.18 What is the purpose of the restriction given in Eq. (6.3.18)?
6.19 Apply the Cveticanin method to

$$\ddot{x} + x = -2\epsilon(\dot{x})^3.$$

Compare this solution to the one obtained from the standard first-order averaging method.
6.20 Can the integral

$$\int \frac{dt}{[a + be^{-ct}]^f},$$

where (a, b, c, f) are constants, be exactly integrated? See Eq. (6.4.35).

References

[1] N. Krylov and N. Bogoliubov, *Introduction to Nonlinear Mechanics* (Princeton University Press; Princeton, NJ; 1943).
[2] N. N. Bogoliubov and J. A. Mitropolsky, *Asymptotical Methods in the Theory of Nonlinear Oscillations* (Hindustan Publishing Co.; Delhi, India; 1963).
[3] R. E. Mickens, *Nonlinear Oscillations* (Cambridge University Press, New York, 1991).
[4] R. E. Mickens and K. Oyedeji, *Journal of Sound and Vibration* **102**, 579 (1985).
[5] R. E. Mickens, *Journal of Sound and Vibration* **264**, 1195 (2003).
[6] L. Cveticanin, *Journal of Sound and Vibration* **320**, 1064 (2008).
[7] R. E. Mickens and S. A. Rucker, *Proceedings of Dynamic Systems and Applications* **4**, 302 (2004).
[8] S. Bravo Yuste and J. Diaz Bejarano, *Journal of Sound and Vibration* **110**, 347 (1986); **139**, 151 (1990); **158**, 267 (1992).
[9] A. H. Nayfeh, *Perturbation Methods* (Wiley, New York, 1973).

[10] R. Bellman, *Perturbation Techniques in Mathematics, Physics and Engineering* (Holt, Rinehart and Winston; New York, 1964).

[11] D. Zwillinger, *Handbook of Differential Equations* (Academic Press, Boston, 1989); see Section 110.

[12] R. E. Mickens, *Mathematical Methods for the Natural and Engineering Sciences* (World Scientific, Singapore, 2004).

[13] R. E. Mickens, K. O. Oyedeji, and S. A. Rucker, *Journal of Sound and Vibration* **268**, 839 (2003).

[14] L. Cveticanin, *Physica A* **341**, 123 (2004).

[15] J. Awrjewitz and I. V. Andrianov, *Journal of Sound and Vibration* **252**, 962 (2002).

[16] H. Hu, Z.-G. Xiong, *Journal of Sound and Vibration* **259**, 977 (2003).

[17] W. T. van Horssen, *Journal of Sound and Vibration* **260**, 961 (2003).

[18] H. P. W. Gottlieb, *Journal of Sound and Vibration* **261**, 557 (2003).

[19] R. E. Mickens, *Mathematical Methods for the Natural and Engineering Sciences* (World Scientific, Singapore, 2004).

[20] B. O. Pierce, *A Short Table of Integrals* (Ginn, Boston, 1929).

[21] P. F. Byrd and M. S. Friedmann, *Handbook of Elliptic Integrals for Engineers and Physicists* (Springer-Verlag, Berlin, 1954).

[22] N. W. McLachlan, *Ordinary Non-Linear Differential Equations in Engineering and Physical Sciences* (Clarendon Press, Oxford, 1956, 2nd edition).

[23] A. H. Nayfeh and D. T. Mook, *Nonlinear Oscillations* (Wiley-Interscience, New York, 1979).

[24] S. H. Strogatz, *Nonlinear Dynamics and Chaos with Applications to Physics, Biology, Chemistry, and Engineering* (Addison-Wesley; Reading, MA; 1994).

Chapter 7

Comparative Analysis

7.1 Purpose

In this final chapter, we consider six of the studied TNL oscillator equations and compare results obtained by all of the methods that were used to calculate approximations to their periodic or oscillatory solutions. Two of the equations are conservative and are given by the following expressions

$$\ddot{x} + x^3 = 0, \tag{7.1.1}$$

$$\ddot{x} + x^{1/3} = 0. \tag{7.1.2}$$

For both equations, we can calculate the exact values of the angular frequencies. Consequently, one measure of the accuracy or quality of a given method is the difference between the exact value of the angular frequency and that determined using the approximation procedure. The initial conditions for Eqs. (7.1.1) and (7.1.2) are taken to be

$$x(0) = A, \quad \dot{x}(0) = 0. \tag{7.1.3}$$

The other four differential equations are non-conservative. Two of the equations have linear damping, i.e.,

$$\ddot{x} + x^3 = -2\epsilon\dot{x}, \tag{7.1.4}$$

$$\ddot{x} + x^{1/3} = -2\epsilon\dot{x}, \tag{7.1.5}$$

while the other two have van der Pol type dissipation, i.e.,

$$\ddot{x} + x^3 = \epsilon(1 - x^2)\dot{x}, \tag{7.1.6}$$

$$\ddot{x} + x^{1/3} = \epsilon(1 - x^2)\dot{x}. \tag{7.1.7}$$

For all four cases, the parameter ϵ is assumed to be small, i.e.,

$$0 < \epsilon \ll 1. \tag{7.1.8}$$

In general, the solutions for Eqs. (7.1.4) and (7.1.5) are expected to oscillate with amplitudes that decrease in magnitude with increasing time. However, the behavior of solutions for Eqs. (7.1.6) and (7.1.7) will depend on the initial conditions since both have stable limit-cycles [1, 2]. Consequently, in the (x, y) phase-space, for initial conditions that lie interior to the closed curve corresponding to the limit-cycle, the trajectories will spiral out to the periodic limit-cycle, while the opposite behavior will occur for initial conditions outside the limit-cycles [1, 2].

For comparison, we consider the approximations to the periodic solutions for the TNL cubic and cube-root conservative oscillators, Eqs. (7.1.1) and (7.1.2), the methods of harmonic balance, parameter expansion and iteration. For the cubic nonconservative oscillators, Eqs. (7.1.4) and (7.1.5), we compare the solutions determined from application of the Mickens-Oyedeji [3], Mickens [4], and Cveticanin [5] procedures. However, for the cube-root, nonconservative oscillators, only the combined linearization-averaging [5] and Cveticanin [5] methods are examined.

This chapter ends with some general comments on TNL oscillators and a list of several unresolved research problems.

7.2 $\ddot{x} + x^3 = 0$

7.2.1 *Harmonic Balance*

The first and second order direct harmonic balance methods give, respectively, the following expressions for the periodic solutions (see Section 3.2.1):

First-Order HB

$$x_1(t) = A\cos(\Omega_1 t), \tag{7.2.1}$$

$$\Omega_1(A) = \left(\frac{3}{4}\right)^{1/2} A. \tag{7.2.2}$$

Second-Order HB

$$x_2(t) = \left(\frac{A}{1+z}\right)[\cos(\Omega_2 t) + z\cos(3\Omega_2 t)], \tag{7.2.3}$$

$$\Omega_2(A) = \left(\frac{3}{4}\right)^{1/2} A\left[\frac{\sqrt{1+z+2z^2}}{1+z}\right] \equiv \Omega_1(A)g(z), \tag{7.2.4}$$

$$z = 0.044818. \tag{7.2.5}$$

The periods derived from these calculations are ($T = 2\pi/\Omega$)

$$T_{\text{exact}}(A) = \frac{7.4163}{A}, \quad T_1(A) = \frac{7.2554}{A}, \quad T_2(A) = \frac{7.4016}{A}, \tag{7.2.6}$$

and they have the following percentage errors

$$\left|\frac{T_{\text{exact}} - T_1}{T_{\text{exact}}}\right| \cdot 100 = 2.2\%, \qquad \left|\frac{T_{\text{exact}} - T_2}{T_{\text{exact}}}\right| \cdot 100 = 0.20\%. \tag{7.2.7}$$

As expected, the higher-order harmonic balance evaluation produces the more accurate estimate for the period or angular frequency. In fact, the second-order harmonic balance has a percentage error smaller than a factor of ten in comparison to the first-order calculation.

Another feature of the direct harmonic balance technique, in particular as applied to this equation, is that the k-th order approximation contains $(3^k + 1)/2$ harmonics. Thus, there is a very rapid increase in the number of algebraic and trigonometrical operations with k. A major difficulty with harmonic balancing, for $k \geq 2$, is the need to solve systems of coupled, nonlinear algebraic equations. However, our calculations clearly show that the second-order harmonic balance results provide an accurate approximation to the periodic solution.

The rational harmonic balance approach (see Section 3.4.1) was applied to the pure-cubic Duffing equation with the following results obtained:

Rational HB

$$x_{\text{RHB}}(t) = \frac{(0.909936)A\cos[\Omega_{\text{RHB}}t]}{1 - (0.090064)\cos[2\Omega_{\text{RHB}}t]}, \tag{7.2.8}$$

$$\Omega_{\text{RHB}}(A) = (0.847134)A, \tag{7.2.9}$$

$$\text{percentage error} = \left|\frac{T_{\text{exact}} - T_{\text{RHB}}}{T_{\text{exact}}}\right| \cdot 100 = 0.01\%. \tag{7.2.10}$$

These results indicate that the rational harmonic balance procedure gives a very good estimate for the angular frequency (or period) and its representation, as shown in Eq. (7.2.8), provides contributions to the periodic solution from all the relevant harmonics.

7.2.2 *Parameter Expansion*

The parameter expansion calculation for the periodic solution is given in Section 4.2.1. To order p, with p set to one in the final results, we found

$$x_{\text{PE}}(t) = A\left[\left(\frac{23}{24}\right)\cos(\Omega_{\text{PE}}t) + \left(\frac{1}{23}\right)\cos(3\Omega_{\text{PE}}t)\right], \tag{7.2.11}$$

$$\Omega_{\text{PE}}(A) = \left(\frac{3}{4}\right)^{1/2} A. \tag{7.2.12}$$

Note that to order p, the angular frequency is the same as that obtained in the first-order harmonic balance calculation, i.e., from Eq. (7.2.2), we have on comparison with Eq. (7.2.12), the result

$$\Omega_{\text{PE}}(A) = \Omega_1(A) = \left(\frac{3}{4}\right)^{1/2} A. \tag{7.2.13}$$

If we further compare Eq. (7.2.3) and Eq. (7.2.11), and observe that

$$z = 0.044818 \approx \frac{1}{23}, \tag{7.2.14}$$

then it follows that the $O(p)$ solution obtained from the parameter expansion method is (essentially) the same as that derived from the second-order harmonic balance procedure, provided we replace Ω_2 by Ω_{PE}.

7.2.3 *Iteration*

Three approximations were calculated using the direct iteration method; see Section 5.2.1:

Zero-Order

$$x_0(t) = A\cos(\Omega_0 t), \quad \Omega_0(A) = \left(\frac{3}{4}\right)^{1/2} A; \tag{7.2.15}$$

First-Order

$$x_1(t) = A\left[\left(\frac{23}{24}\right)\cos(\Omega_0 t) + \left(\frac{1}{23}\right)\cos(3\Omega_0 t)\right]; \tag{7.2.16}$$

Second-Order

$$x_2(t) = A\Big\{(0.955)\cos\theta + (4.29)\cdot 10^{-2}\cos 3\theta + (1.73)\cdot 10^{-3}\cos 5\theta + (3.60)\cdot 10^{-5}\cos 7\theta + (3.13)\cdot 10^{-7}\cos 9\theta\Big\}, \tag{7.2.17}$$

$$\theta = \Omega_1(A)t, \tag{7.2.18}$$

where $\Omega_{\text{exact}}(A)$, $\Omega_0(A)$, and $\Omega_1(A)$ are listed below with their respective percentage errors:

$$\begin{cases} \Omega_{\text{exact}}(A) = (0.84723)A, & \\ \Omega_0(A) = (0.866025)A, & \% \text{ error} = 2.2, \\ \Omega_1(A) = (0.849326)A, & \% \text{ error} = 0.2. \end{cases} \tag{7.2.19}$$

A calculation was also performed by application of the extended iteration method; see Section 5.3.1. The results obtained for this case are:

Extended Iteration

- $x_0(t)$ and $x_1(t)$ are exactly the expressions given by Eqs. (7.2.15) and (7.2.16).
- $x_2(t)$ is now

$$x_2(t) = \left(\frac{13,244}{12,672}\right)\cos\theta - \left(\frac{595}{12,672}\right)\cos 3\theta + \left(\frac{23}{12,672}\right)\cos 5\theta, \tag{7.2.20}$$

$$\theta = \Omega_{\text{EI}}(A)t, \quad \Omega_{\text{EI}}(A) = \left(\frac{33}{46}\right)^{1/2} A = (0.84699)A, \tag{7.2.21}$$

and the percentage error for $\Omega_{\text{EI}}(A)$ is 0.03%.

A comparison of the direct and extended iteration procedures leads to the following general conclusions:

(i) At the k-th level of iteration, the direct iteration procedure produces a solution $x_k(t)$ containing $(3^k + 1)/2$ harmonics, while the extended iteration method gives a solution having only $k + 1$ harmonics. Therefore, from the standpoint of computational effort, extended iteration has an advantage in comparison to the direct iteration method.

(ii) For either iteration procedure, the magnitude of the coefficients for the higher harmonics decrease rapidly; in fact, their decrease is consistent with an exponential fall off in values.

(iii) The extended iteration procedure, at the $k = 2$ level, gives the better estimate for the angular frequency, i.e., 0.03 percentage error versus 0.2 percentage error.

7.2.4 *Comments*

Three procedures, along with refinements, were applied to the calculation of approximations to the periodic solutions and the associated angular frequencies for the purely cubic Duffing equation

$$\ddot{x} + x^3 = 0.$$

The two methods producing the more accurate solutions were the rational harmonic balance (Section 3.4.1) and the extended iteration (Section 5.3.1) techniques. They have the additional advantage, in comparison to the other procedures, of being computationally efficient.

7.3 $\ddot{x} + x^{1/3} = 0$

7.3.1 *Harmonic Balance*

The cube-root TNL oscillator

$$\ddot{x} + x^{1/3} = 0, \tag{7.3.1}$$

was studied using two different first-order harmonic balance procedures. For the first calculation, Eq. (7.3.1) was used, while for the second, the following expression was employed (see Section 3.2.4)

$$(\ddot{x})^3 + x = 0. \tag{7.3.2}$$

The corresponding solutions, angular frequencies, and percentage errors found were

$$\text{Eq. (7.3.1)}: \quad \begin{cases} x^{(1)}_{\text{HB}\,1}(t) = A\cos[\Omega^{(1)}_1 t], \\ \Omega^{(1)}_1 = \dfrac{1.076844}{A^{1/3}}, \\ \%\text{ error in } \Omega^{(1)}_1 = 0.6\%; \end{cases} \tag{7.3.3}$$

$$\text{Eq. (7.3.2)}: \quad \begin{cases} x^{(2)}_{\text{HB}\,2}(t) = A\cos[\Omega^{(2)}_1 t], \\ \Omega^{(2)}_1 = \dfrac{1.049115}{A^{1/3}}, \\ \%\text{ error in } \Omega^{(2)}_1 = 2.0\%; \end{cases} \tag{7.3.4}$$

where

$$\Omega_{\text{exact}}(A) = \frac{1.070451}{A^{1/3}}. \tag{7.3.5}$$

Second-order harmonic balance can only be applied to Eq. (7.3.2), and carrying out this procedure gives

$$\begin{cases} x_{\text{HB}\,2}(t) = \left(\dfrac{A}{1+z}\right)[\cos\theta + z\cos 3\theta] \\ \theta = \Omega_2(A)t, \quad \Omega_2(A) = \dfrac{1.063410}{A^{1/3}}, \\ \%\text{ error in } \Omega_2(A) = 0.7\%. \\ \qquad z = -0.019178 \approx \dfrac{1}{52}. \end{cases} \tag{7.3.6}$$

Clearly, the second-order harmonic balance produces a very accurate value for the angular frequency. Note that the coefficient of the third harmonic is only about 2% of the mangitude of the first-harmonic.

To obtain the results, for the second-order harmonic balance application, two coupled, cubic algebraic equations had to be solved. Since one of the equations was homogeneous, of degree two in the variables, this made finding the required solutions easier.

7.3.2 *Parameter Expansion*

Section 4.2.4 contains the calculations for an order-p determination of the solution to the cube-root equation. We found the following results

$$\begin{cases} x_{\text{PE}}(t) = \left(\dfrac{25}{24}\right) A \left[\cos\theta - \left(\dfrac{1}{25}\right)\cos 3\theta\right], \\ \theta = \Omega_{\text{PE}}t, \quad \Omega_{\text{PE}}(A) = \left(\dfrac{4}{3}\right)^{1/6} \dfrac{1}{A^{1/3}} = \dfrac{1.070451}{A^{1/3}}, \\ \%\text{ error in } \Omega_{\text{PE}}(A) = 2\%. \end{cases} \tag{7.3.7}$$

These quantities were determined from the use of the following p-method reformulation of the cube-root equation:

$$\ddot{x} + \Omega^2 x = p[\ddot{x} - \Omega^2(\ddot{x})^3]. \tag{7.3.8}$$

If however, we use

$$0 \cdot \ddot{x} + 1 \cdot x = -p(\ddot{x})^3, \tag{7.3.9}$$

then a completely different, "unphysical solution" is found; see Section 4.2.4, Eq. (4.2.57). This finding implies that we need to be very careful in the formulation of TNL differential equations for which the p-expansion will be applied.

7.3.3 *Iteration*

The zero- and first-order iteration solutions and angular frequencies, based on the method of direct iteration are (see Section 5.2.5):

First-Order Iteration

$$x_0(t) = A\cos(\Omega_0 t), \tag{7.3.10}$$

$$\Omega_0(A) = \left(\frac{4}{3}\right)^{1/6} \frac{1}{A^{1/3}} = \frac{1.0491151}{A^{1/3}}, \tag{7.3.11}$$

$$\% \text{ error in } \Omega_0(A) = 2.0\%. \tag{7.3.12}$$

Second-Order Iteration

$$x_1(t) = \left(\frac{25}{24}\right) A \left[\cos\theta - \left(\frac{1}{25}\right)\cos 3\theta\right], \tag{7.3.13}$$

$$\theta = \Omega_1 t, \quad \Omega_1(A) = \frac{1.041427}{A^{1/3}}, \tag{7.3.14}$$

$$\% \text{ error in } \Omega_1(A) = 2.7\%. \tag{7.3.15}$$

Several issues should be noted from these calculations. First, the value of the angular frequency is slightly better for the first-iteration solution in comparison to that found by the second-iteration. This result may indicate, for this particular TNL oscillator, that higher-order iterative solutions may not be reliable. Second, the different methods of calculating approximations to the periodic solutions of TNL oscillator equations may give exactly the same results for the solution functions, but provide different estimates of the angular frequencies. In particular, see the first of Eqs. (7.3.7) and Eq. (7.3.13), and the second of Eqs. (7.3.7) and Eq. (7.3.14).

7.3.4 *Comment*

The discrepancies between the three methods used to approximate the periodic solutions of the cube-root equation may occur because $x^{1/3}$ is not analytic at $x = 0$. This means, in particular, that such derivatives do not exist for $d^n x/dt^n$, if $n \geq 3$. Overall, the harmonic balance method appears to be the better procedure in comparison with both parameter expansion and iteration methods.

7.4 $\ddot{x} + x^3 = -2\epsilon\dot{x}$

We now study a linearly damped TNL oscillator. This equation is

$$\ddot{x} + x^3 = -2\epsilon\dot{x}, \quad 0 < \epsilon \ll 1, \tag{7.4.1}$$

and corresponds to a linearly damped, pure cubic, Duffing differential equation. We examine the approximations to the solutions using both the Mickens-Oyedeji [3] and the Cveticanin [5] procedures. Note that both methodologies are based on the requirement that the parameter ϵ is very small [3, 5, 6]. A brief discussion of the calculations derived from the combined linearization-averaging technique is also provided.

7.4.1 *Mickens-Oyedeji*

The approximation to the damped oscillatory solution to Eq. (7.4.1) takes the form (see Section 6.2.1)

$$x(t) \simeq a(t,\epsilon)\cos\psi(t,\epsilon), \tag{7.4.2}$$

where

$$a(t,\epsilon) = Ae^{-\epsilon t}, \tag{7.4.3}$$

$$\psi(t,\epsilon) = \left(\frac{\sqrt{3}}{4}\right)At + \left(\frac{\sqrt{3}}{4}\right)A\left[\frac{1-e^{-2\epsilon t}}{2\epsilon}\right]. \tag{7.4.4}$$

To obtain these results, the initial conditions were selected to be

$$a(0,\epsilon) = A, \quad \psi(0,\epsilon) = 0. \tag{7.4.5}$$

If we define a time-dependent angular frequency as

$$\Omega(t,A,\epsilon) \equiv \frac{\psi(t,A,\epsilon)}{t}, \tag{7.4.6}$$

where $\psi(t,A,\epsilon)$ is the function in Eq. (7.4.4), then

$$\Omega(t,A,\epsilon) = \left[\frac{\Omega_0(A)}{2}\right]\left\{1 + \left[\frac{1-e^{-2\epsilon t}}{2\epsilon t}\right]\right\}, \tag{7.4.7}$$

where

$$\Omega_0(A) = \left(\frac{3}{4}\right)^{1/2} A, \tag{7.4.8}$$

is the approximation to the angular frequency when $\epsilon = 0$, i.e., it is the value calculated using harmonic balance for the equation

$$\ddot{x} + x^3 = 0. \tag{7.4.9}$$

From Eq. (7.4.7), it follows that $\Omega(t, A, \epsilon)$ has the properties

$$\Omega(t, A, \epsilon)_{t \text{ small}} = \Omega_0(A) - \left[\frac{\Omega_0(A)\epsilon}{2}\right] t + O(t^2), \tag{7.4.10}$$

$$\Omega(t, A, \epsilon)_{t \text{ large}} \sim \frac{\Omega_0(A)}{2} + \left[\frac{\Omega_0(A)}{4\epsilon}\right]\left(\frac{1}{t}\right). \tag{7.4.11}$$

Figure 7.4.1 gives a plot of $\Omega(t, A, \epsilon)$ versus t, for fixed A and ϵ.

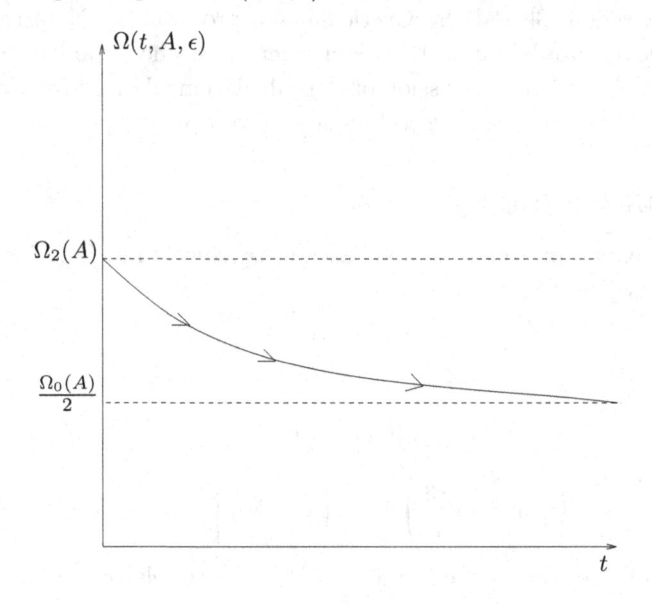

Fig. 7.4.1 Plot of $\Omega(t, A, \epsilon)$ versus t, for the linearly damped, pure cubic Duffing equation. $\Omega_0(A) = \left(\frac{3}{4}\right)^{1/2} A$.

One of the important predictions coming from Eq. (7.4.7) is that the apparent period of the damped, oscillatory motion should increase with time. To see this, define the period $T(t, A, \epsilon)$ to be

$$T(t, A, \epsilon) = \frac{2\pi}{\Omega(t, A, \epsilon)}. \tag{7.4.12}$$

Since $\Omega(T, A, \epsilon)$ decreases from $\Omega_0(A)$ at $t = 0$, to the value $\Omega_0(A)/2$ for large times, it follows that $T(t, A, \epsilon)$ is an increasing function of the time.

Figure 7.4.2 gives a plot of $x(t)$ versus t. Two important features of this graph are the smooth decrease of the amplitude and the increase of the period of the oscillations with increasing time. Both properties are consistent with the predicted features coming from the calculations based on the Mickens-Oyedeji method [3].

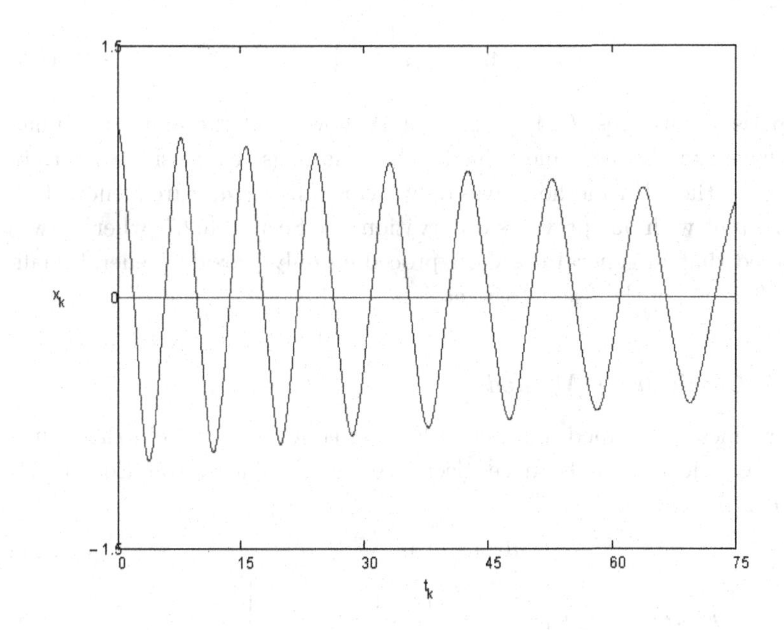

Fig. 7.4.2 Plot of the numerical solution of $\ddot{x} + x^3 = -2\epsilon\dot{x}$ for $\epsilon = 0.01$, $x(0) = 1$ and $\dot{x}(0) = 0$.

7.4.2 *Combined-Linearization-Averaging*

The combined-linearization-averaging (CLA) method replaces Eq. (7.4.1)

$$\ddot{x} + x^3 = -2\epsilon x, \quad 0 < \epsilon \ll 1,$$

by

$$\ddot{x} + [\Omega_0(A)]^2 x = -2\epsilon\dot{x}, \tag{7.4.13}$$

where

$$\Omega_0(A) = \left(\frac{3}{4}\right)^{1/2} A. \tag{7.4.14}$$

Equation (7.4.13) is solved under the assumption that its solution fulfills the following initial conditions

$$x(0) = A, \quad \dot{x}(0) = 0. \tag{7.4.15}$$

Since Eq. (7.4.13) is a linear differential equation, its solution can be easily calculated and is found to be (to terms of order ϵ; see Eqs. (6.2.29) to (6.2.31))

$$x(t,\epsilon) \simeq Ae^{-\epsilon t}\cos\left[\left(\frac{3}{4}\right)^{1/2} At\right]. \tag{7.4.16}$$

Comparison with Eqs. (7.4.3) and (7.4.4) shows that the amplitude functions have exactly the same behavior, but the phase expressions differ. In particular, the CLA method gives only a constant angular frequency. This is consistent with our previous observation, see Section 6.2.7, where it was remarked that in general the CLA procedure only provides general qualitative features of the oscillatory motion.

7.4.3 *Cveticanin's Method*

For the linearly damped, pure cube Duffing equation, $\alpha = 3$ in the framework of Cveticanin's method; see Section 6.4.1. Therefore, from Eqs. (6.4.4) and (6.4.8), we have

$$a(t,\epsilon) = Ae^{-\left(\frac{2\epsilon}{3}\right)t}, \tag{7.4.17}$$

$$\psi(t,\epsilon) = \left(\frac{3}{2}\right)\left[\frac{\sqrt{2\pi}\,\Gamma\left(\frac{3}{4}\right)}{\Gamma\left(\frac{1}{4}\right)}\right]\left[\frac{1-e^{-\left(\frac{2\epsilon}{3}\right)t}}{\epsilon}\right]A. \tag{7.4.18}$$

Note that

$$\lim_{\epsilon\to 0}\psi(t,\epsilon) = \left[\frac{\sqrt{2\pi}\,\Gamma\left(\frac{3}{4}\right)}{\Gamma\left(\frac{1}{4}\right)}\right]At \equiv \Omega(A)t = (0.84721)At, \tag{7.4.19}$$

and, also

$$\underset{t\ \text{small}}{\psi(t,\epsilon)} = (0.84721)At. \tag{7.4.20}$$

If the effective, time-dependent angular frequency is defined as

$$\Omega(t,A,\epsilon) = \frac{\psi(t,A,\epsilon)}{t} = \Omega(A)\left(\frac{3}{2\epsilon}\right)\left[\frac{1-e^{-\left(\frac{2\epsilon}{3}\right)t}}{t}\right], \tag{7.4.21}$$

then $\Omega(t,A,\epsilon)$ has the properties

$$\begin{cases} \Omega(t,A,0) = \Omega(A), \\ \Omega(t,A,\epsilon) > 0 \quad \text{for } t > 0, \\ \underset{t\ \text{large}}{\Omega(t,A,\epsilon)} \sim \Omega(A)\left(\frac{3}{2\epsilon}\right)\left(\frac{1}{t}\right), \\ \dfrac{d\Omega(t,A,\epsilon)}{dt} < 0, \end{cases} \tag{7.4.22}$$

and it follows that $\Omega(t, A, \epsilon)$ monotonically decreases to zero as $t \to \infty$, from the value $\Omega(0, A, \epsilon) = \Omega(A)$ at $t = 0$. One consequence of this result is that the corresponding, time-dependent effective period, see Eq. (7.4.12), monotonically increases as $t \to \infty$. Figure 7.4.2 illustrates this prediction.

7.4.4 *Discussion*

The Mickens-Oyedeji and Cveticanin averaging methods both give the same general properties of the solutions to the linearly damped, pure cubic-Duffing equation. However, they make different predictions with respect to the detailed time dependencies of the amplitude and phase. Table 7.4.1 summarizes these distinctions.

The Mickens combined-linearization-averaging technique only provides the correct qualitative features of the damped oscillations.

Table 7.4.1 Comparison of the amplitude and effective angular frequencies for the linearly damped, pure cubic, Duffing oscillator.

	$a(t, A, \epsilon)$	$\Omega(t, A, \epsilon)$
Mickens-Oyedeji*	$Ae^{-\epsilon t}$	$\left[\frac{\Omega(A)}{2}\right]\left\{1+\left[\frac{1-e^{-2\epsilon t}}{2\epsilon t}\right]\right\}$
Cveticanin**	$Ae^{-\left(\frac{2\epsilon}{3}\right)t}$	$\Omega(A)\left[\frac{1-e^{-\left(\frac{2\epsilon}{3}\right)t}}{\left(\frac{2\epsilon}{3}\right)t}\right]$

*See Eqs. (7.4.3) and (7.4.5).
**See Eqs. (7.4.17) and (7.4.21).
$\Omega(A) = \frac{\sqrt{2\pi}\,\Gamma\left(\frac{3}{4}\right)}{\Gamma\left(\frac{1}{4}\right)}$

7.5 $\ddot{x} + x^{1/3} = -2\epsilon\dot{x}$

7.5.1 *Combined-Linearization-Averaging*

This method gives the following approximation for the linearly damped, cube-root oscillator

$$x(t, \epsilon) \simeq Ae^{-\epsilon t}\cos[\Omega(A)t], \tag{7.5.1}$$

$$\Omega^2(A) = \frac{a_1}{A^{2/3}}, \tag{7.5.2}$$

where $a_1 = 1.1595952669\ldots$. From Eq. (7.5.1) it follows that the amplitude decreases exponentially to zero and the phase, i.e., $\psi(t,\epsilon) = \Omega(A)t$, is a linear function of time.

Figure 7.5.1 presents a plot of the numerical solution for

$$\ddot{x} + x^{1/3} = -2\epsilon\dot{x},$$

for $\epsilon = 0.01$, $x(0) = 1$ and $\dot{x}(0) = 0$. Examination of the graph indicates that it exhibits all the qualitative properties of the function given in Eq. (7.5.1)

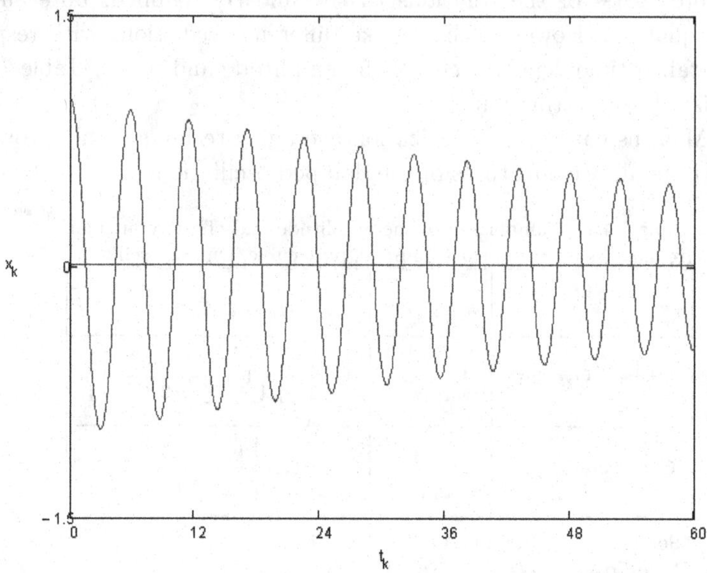

Fig. 7.5.1 Plot of the numerical solution of $\ddot{x} + x^{1/3} = -2\epsilon\dot{x}$ for $\epsilon = 0.01$, $x(0) = 1$ and $\dot{x}(0) = 0$.

7.5.2 *Cveticanin's Method*

Applying Cveticanin's method to the case $\alpha = \frac{1}{3}$, we find the following expressions, respectively, for the amplitude and phase:

$$a(t,\epsilon) = A\exp\left[-\left(\frac{6\epsilon t}{5}\right)\right], \tag{7.5.3}$$

$$\psi(t,\epsilon) = \Omega(A)\left[\frac{\exp\left(\frac{2\epsilon t}{5}\right) - 1}{\left(\frac{2\epsilon}{5}\right)}\right], \tag{7.5.4}$$

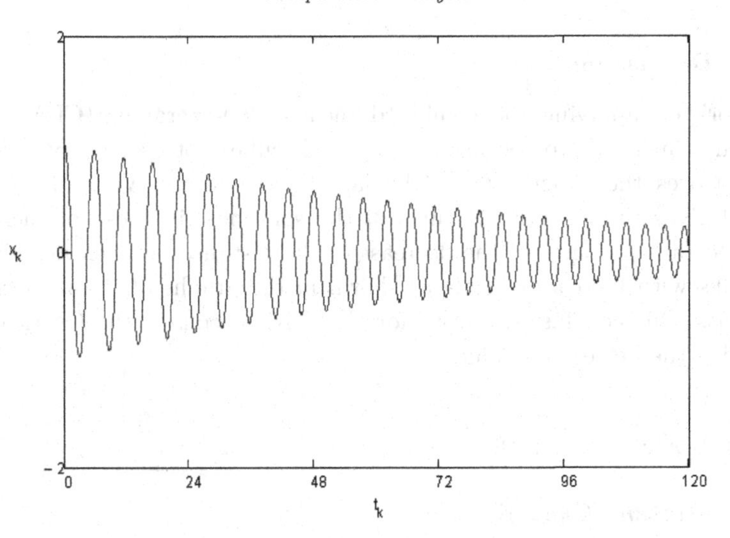

Fig. 7.5.2 This graph is the same as that in Figure 7.5.1, except that the interval in time is twice as long.

where

$$\Omega(A) = \left(\frac{\pi}{6}\right)^{1/2} \left[\frac{\Gamma\left(\frac{1}{4}\right)}{\Gamma\left(\frac{3}{4}\right)}\right] \left(\frac{1}{A^{1/3}}\right) = \frac{1.0768}{A^{1/3}}\,. \tag{7.5.5}$$

Observe that the phase function, $\psi(t,\epsilon)$, is an exponentially increasing function of time.

Denoting $\psi(t,A,\epsilon) = \psi(t,\epsilon)$, then the effective angular frequency is defined to be

$$\Omega(t,A,\epsilon) \equiv \frac{\psi(t,A,\epsilon)}{t} = \Omega(A)\left[\frac{\exp\left(\frac{2\epsilon t}{5}\right)-1}{\left(\frac{2\epsilon}{5}\right)t}\right]. \tag{7.5.6}$$

Similarly, the effective period is

$$T(t,A,\epsilon) \equiv \frac{2\pi}{\psi(t,A,\epsilon)} = \left[\frac{2\pi}{\Omega(A)}\right]\left[\frac{\left(\frac{2\epsilon}{5}\right)t}{\exp\left(\frac{2\epsilon t}{5}\right)-1}\right]. \tag{7.5.7}$$

One consequence of this last formula is that $T(t,A,\epsilon)$ will decrease from $2\pi/\Omega(A)$ at $t=0$, to the value zero as $t \to \infty$. Another result is that the distance between neighborhood peaks of the oscillatory motion will also decrease with time. Figure 7.5.2 illustrates both phenomena.

7.5.3 *Discussion*

Our work on applying the combined-linearization-averaging (CLA) and Cveticanin methods to the linearly damped, cube-root oscillatory clearly demonstrates the superiority of the latter procedure. While the CLA method gives the essential features of the solutions, it does not include the important property of the decrease of the distance between neighboring peaks with increase of time as shown in the graphs of the numerical solutions. Our conclusion is that for this TNL oscillator the Cveticanin method is the better technique.

7.6 $\ddot{x} + x^3 = \epsilon(1 - x^2)\dot{x}$

7.6.1 *Mickens-Oyedeji*

For the pure cubic Duffing type van der Pol equation

$$\ddot{x} + x^3 = \epsilon(1 - x^2)\dot{x}, \quad 0 < \epsilon \ll 1, \tag{7.6.1}$$

the amplitude based on the Mickens-Oyedeji procedure (see Eq. (6.2.23)), is

$$a(t, \epsilon) = \frac{2A}{[A^2 + (4 - A^2)e^{-\epsilon t}]^{1/2}}. \tag{7.6.2}$$

Note that

$$\operatorname*{Lim}_{t \to \infty} a(t, \epsilon) = 2, \tag{7.6.3}$$

a result that holds for any value of $x(0) = A$. This fact implies that Eq. (7.6.1) has a limit-cycle solution such that regardless of the initial conditions the amplitude asymptotically approaches the value $a_\infty = 2$.

The corresponding expression for the phase (see Eqs. (6.1.2) and (6.2.25)) is

$$\begin{aligned}\psi(t, \epsilon) = \left(\frac{\sqrt{3}}{4}\right) At + \left(\frac{\sqrt{3}A}{4}\right) \Bigg\{ \left(\frac{4 - A^2}{\epsilon A^2}\right) \ln\left[(4 - A^2) + A^2 e^{\epsilon t}\right] \\ + \ln\left[A^2 + (4 - A^2)e^{-\epsilon t}\right] \Bigg\} - \left[\frac{\sqrt{3}\ln(4)}{4\epsilon A}\right] \left[4 - (1 - \epsilon)A^2\right].\end{aligned} \tag{7.6.4}$$

This relation for $\psi(t, \epsilon)$ has the property that the phase depends on the initial value, $x(0) = A$, for all $t > 0$. However, a defining feature of limit-cycles is that the associated phase, as $t \to \infty$, is independent of the initial conditions. Therefore, we must conclude that the Mickens-Oyedeji method does not provide an appropriate solution for Eq. (7.6.1),

$$x(t) \simeq a(t, \epsilon) \cos \psi(t, \epsilon), \tag{7.6.5}$$

if $a(t, \epsilon)$ and $\psi(t, \epsilon)$ are given by Eqs. (7.6.2) and (7.6.4).

7.6.2 *Cveticanin's Method*

From Section 6.4.3, with $\alpha = 3$, we find the following expression for the amplitude function

$$a(t, \epsilon) = \frac{2A}{\left\{A^2 + (4 - A^2) \exp\left[-\left(\frac{2\epsilon}{3}\right) t\right]\right\}^{1/2}}, \tag{7.6.6}$$

and $a(t, \epsilon)$ has the property

$$\operatorname*{Lim}_{t \to \infty} a(t, \epsilon) = 2. \tag{7.6.7}$$

The derivative of the phase function is

$$\dot{\psi} = \frac{2qA}{\left\{A^2 + (4 - A^2) \exp\left[-\left(\frac{2\epsilon}{3}\right) t\right]\right\}^{1/2}}. \tag{7.6.8}$$

For large t, $\psi(t, \epsilon)$ is given by

$$\psi(t, \epsilon) \sim (2q)t. \tag{7.6.9}$$

Using

$$q(\alpha = 3) = \frac{\sqrt{2\pi}\, \Gamma\left(\frac{3}{4}\right)}{\Gamma\left(\frac{1}{4}\right)} = 0.8477213, \tag{7.6.10}$$

we calculate $\psi(t, \epsilon)$ to be

$$\psi(t, \epsilon) \sim (1.7320508)t. \tag{7.6.11}$$

Therefore, using Cveticanin's procedure, we obtain

$$x(t, \epsilon) \underset{\text{large t}}{\longrightarrow} 2 \cos(1.6954426)t. \tag{7.6.12}$$

7.6.3 *Discussion*

If first-order harmonic balance is applied to Eq. (7.6.1), we find

$$x_{\mathrm{HB}}(t) = 2\cos(\sqrt{3}\,t) = 2\cos(1.7320508)t. \tag{7.6.13}$$

This follows from the fact that the harmonic balance angular frequency is

$$\Omega_{\mathrm{HB}}(A)\big|_{A=2} = \left(\frac{\sqrt{3}}{2}\right) A\big|_{A=2} = \sqrt{3}. \tag{7.6.14}$$

Thus, the percentage error between the angular frequencies from the Cveticanin method, Eq. (7.6.11), and harmonic balance, Eq. (7.6.13), is about 2%.

In summary, the Cveticanin procedure is the appropriate averaging method to apply to the pure cubic Duffing type van der Pol equation. Of the three averaging methods (Mickens-Oyedeji [3], Mickens [4], and Cveticanin [5]), it is the only one to produce all of the expected features of the limit-cycle solution.

Figures 7.6.1 and 7.6.2 give numerical solutions of the cubic Duffing type van der Pol oscillator, i.e., Eq. (7.6.2), for two different sets of initial conditions. The graphs of Figure 7.6.1 correspond to $x(0) = 4$, $y(0) = dx(0)/dt = 0$, and $\epsilon = 0.1$; while for Figure 7.6.2, we have $x(0) = 0.1$, $y(0) = 0$, and $\epsilon = 0.1$.

In Figure 7.6.1, the initial condition, $x(0) = 4$, is larger than the value of the limit-cycle amplitude. Therefore, $x(t)$ oscillates with decreasing amplitude down to the limit-cycle value. The opposite situation occurs in Figure 7.6.2. For this case the initial condition is smaller than the limit-cycle amplitude and $x(t)$ oscillates with increasing amplitude, approaching the limit-cycle behavior from below. Note that Figure 7.6.2 clearly illustrates the time dependent nature of the angular frequency.

Inspection of both Figures 7.6.1 and 7.6.2 shows that the limiting value of the amplitude determined from the numerical solution is less than two, i.e.,

$$\operatorname*{Lim}_{t\to\infty} a(t,\epsilon) < 2.$$

However, the predicted value from the Cveticanin procedure is two. The resolution of this issue is based on the fact that the first-order averaging method only provides estimates for the amplitude and phase up to terms of order ϵ [1, 6, 8].

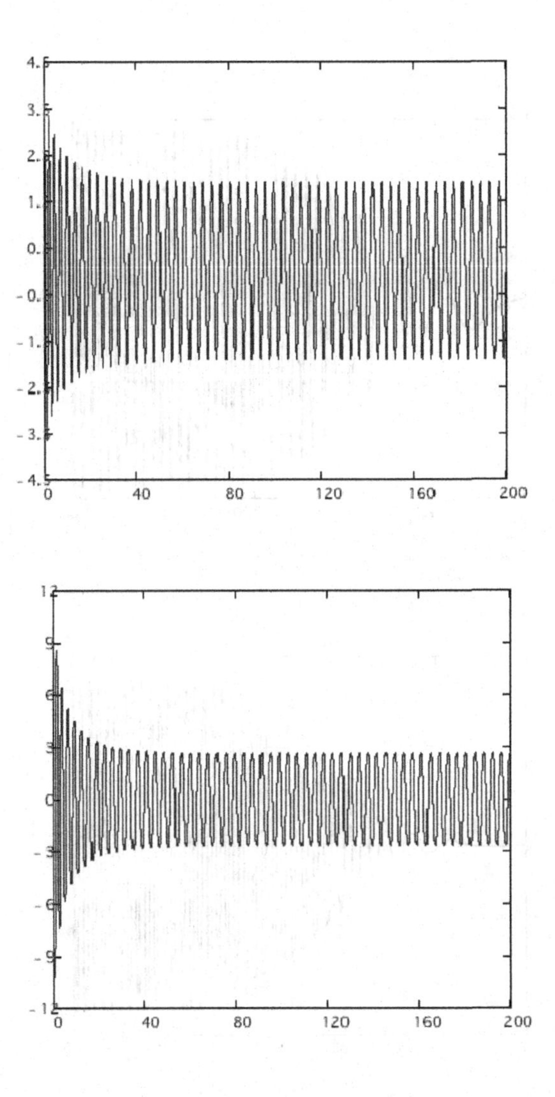

Fig. 7.6.1 Numerical solution of Eq. (7.6.1) for $x(0) = 4$, $y(0) = 0$, and $\epsilon = 0.1$.

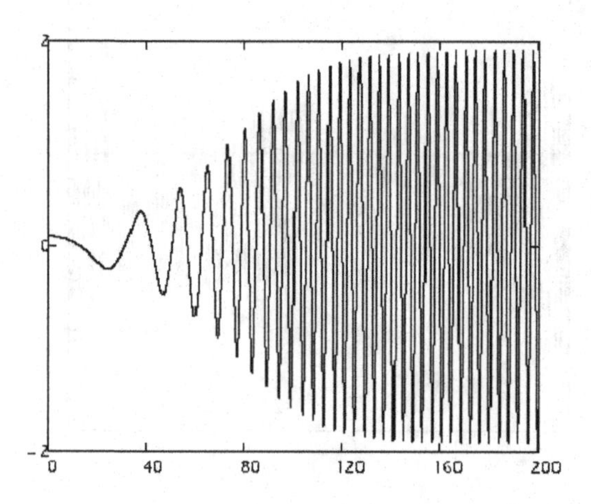

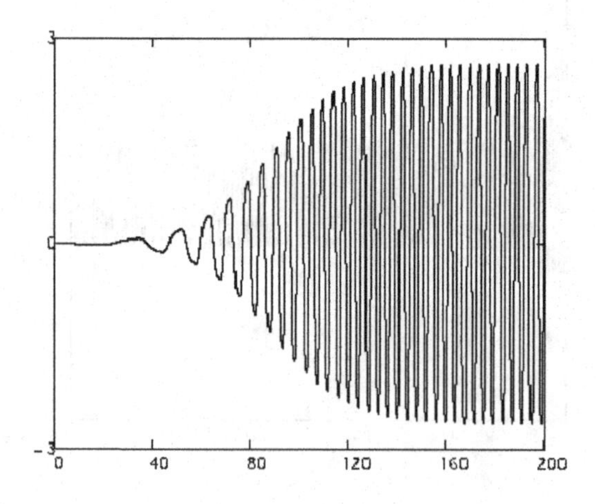

Fig. 7.6.2 Numerical solution of Eq. (7.6.2) for $x(0) = 0.1$, $y(0) = 0$, and $\epsilon = 0.1$.

7.7 $\ddot{x} + x^{1/3} = \epsilon(1 - x^2)\dot{x}$

Since the combined-linearization-averaging (CLA) method [4] only provides a qualitative description of the general features of solutions to dissipative TNL oscillators, we will not discuss it as applied to the above differential equation. However, within the framework of the Cveticanin procedure [5], see also Section 6.4.3, we have for

$$\ddot{x} + x^{1/3} = \epsilon(1 - x^2)\dot{x}, \quad 0 < \epsilon \ll 1, \tag{7.7.1}$$

the following expression for the amplitude

$$a(t, \epsilon) = \frac{2A}{\left\{A^2 + (4 - A^2)\exp\left[-\left(\frac{6\epsilon}{5}\right)t\right]\right\}^{1/2}}. \tag{7.7.2}$$

This function has the property

$$\lim_{t \to \infty} a(t, \epsilon) = 2. \tag{7.7.3}$$

The derivative of the phase is

$$\dot{\psi}(t, \epsilon) = \left[\frac{q}{(2A)^{1/3}}\right]\left\{A^2 + (4 - A^2)\exp\left[-\left(\frac{6\epsilon}{5}\right)t\right]\right\}^{1/6}, \tag{7.7.4}$$

and for large t it becomes

$$\dot{\psi}(t, \epsilon) \sim \left(\frac{q}{2^{1/3}}\right) = \left(\frac{1}{2^{1/3}}\right)\left(\frac{\pi}{24}\right)^{1/2}\left[\frac{\Gamma\left(\frac{1}{4}\right)}{\Gamma\left(\frac{3}{4}\right)}\right]. \tag{7.7.5}$$

One consequence of the last expression is that

$$\psi(t, \epsilon) \sim \left(\frac{q}{2^{1/3}}\right)t = (0.849617)t. \tag{7.7.6}$$

The results from Eqs. (7.7.3) and (7.7.6) imply that the cube-root van der Pol equation has a limit cycle solution such that for arbitrary initial conditions all solutions approach a closed path of the $(x, y = \dot{x})$ phase-plane having an amplitude of value two and an angular frequency given by the coefficient of t in Eq. (7.7.6). These results are in agreement with those determined by other methods [7].

7.8 General Comments and Calculation Strategies

Sections 7.2 to 7.7 have given brief overviews of the results from determining approximations to the oscillatory solutions for six model TNL oscillators. This section provides a general summary of these findings and suggest a strategy for carrying out calculations on this class of nonlinear oscillator differential equations.

7.8.1 *General Comments*

(A) Given a particular TNL oscillator differential equation, the first goal is to show that it has periodic solutions. Five possible cases may occur. The following is a listing of these cases, along with a representative TNL equation

1) All solutions are periodic,

$$\ddot{x} + x^3 = 0. \tag{7.8.1}$$

2) Some solutions are periodic,

$$\ddot{x} + (1 + \dot{x})x^{1/3} = 0. \tag{7.8.2}$$

3) Oscillatory solutions exist,

$$\ddot{x} + x = -2\epsilon\dot{x}. \tag{7.8.3}$$

4) A limit-cycle occurs,

$$\ddot{x} + x^{1/3} = \epsilon(1 - x^2)\dot{x}. \tag{7.8.4}$$

5) No oscillatory or periodic solutions exist,

$$\ddot{x} - x^3 = 0. \tag{7.8.5}$$

For most cases, the use of phase-space analysis will provide the required proof or non-proof of either periodic or oscillatory motions; see Chapter 2.

(B) The application of a particular calculational method to determine solutions must be preceded by the transformation of the original TNL differential equation into one appropriate for the method. For example, the equation

$$\ddot{x} + x^{1/3} = 0,$$

cannot be used for a second-order harmonic balance analysis. It must first be placed in the form

$$(\ddot{x})^3 + x = 0,$$

before the harmonic balance method can be applied. Likewise, for an iteration calculation, this same equation must be rewritten to the expression

$$\ddot{x} + \Omega^2 x = \ddot{x} - \Omega^2(\ddot{x})^3,$$

which then becomes

$$\ddot{x}_{k+1} + \Omega_k^2 x_k = \ddot{x}_k - \Omega_k^2(\ddot{x}_k)^3.$$

(C) For a conservative TNL oscillator, i.e.,

$$\ddot{x} + g(x) = 0,$$

the methods of harmonic balance, parameter expansion, and iteration may be applied. If the exact angular frequency is known, then after an approximation expression is obtained for the solution, the approximate function for the angular frequency can be replaced by the exact value of the angular frequency.

To illustrate this, suppose that we only wish to obtain a first-order solution and it is given by

$$x_1(t) = \left(\frac{A}{1+z}\right)[\cos\theta + z\cos 3\theta],$$

where z is known and

$$\theta \equiv \Omega_1(A)t.$$

If $\Omega_e(A)$ is the exact angular frequency, then the modified solution is

$$x_1(t) \to x_1^{(m)}(t) = \left(\frac{A}{1+z}\right)[\cos\theta_m + z\cos 3\theta_m]$$

where

$$\theta_m = \Omega_e(A)t.$$

It is expected that $x_1^{(m)}(t)$ will provide a better approximation to the actual solution than $x_1(t)$.

(D) The extended iteration method should be used rather than the direct procedures; see Section 7.5.1. In general, the extended iteration method, for a given level of calculation, gives a more accurate value for the angular frequency and is computationally less demanding than the direct procedure.

(E) For nonconservative, dissipative TNL oscillators, only the Cveticanin method should be applied to determine accurate approximations to the oscillatory solutions.

7.8.2 *Calculation Strategies*

The investigations presented in Chapters 3, 4, 5 and 6 provide guidance as to how to proceed in the process of calculating periodic and/or oscillatory solutions for TNL oscillator differential equations. There are two cases to consider.

Conservative Oscillators

For this situation, the equation of motion is

$$\ddot{x} + g(x) = 0, \quad x(0) = A, \quad \dot{x}(0) = 0 \tag{7.8.6}$$

where $g(x)$ is of odd parity, i.e.,

$$g(-x) = -g(x), \tag{7.8.7}$$

and we assume that $g(x)$ is such that all solutions are periodic.

- First, carry out a preliminary analysis using first-order harmonic balance. This analysis will provide an overall view of how the angular frequency depends on the amplitude A. It may also alert you to possible difficulties that exist in either higher order harmonic balance procedures or in other methods, such as parameter expansion and iteration.
- Second, determine if an exact, closed form expression can be calculated for the angular frequency. In general, it is not likely that the angular frequency function exists in a form expressible in terms of a finite number of the known standard functions.
- Third, attempt the calculation of a second-order harmonic balance solution. This procedure will lead to two, coupled, very nonlinear, algebraic equations. If these equations can be either exactly or approximately solved, then a satisfactory solution can be found.
- If a higher level solution, in terms of included harmonics, is required, then use an extended iteration method to calculate these approximations.

Nonconservative Oscillators

For this case, the TNL oscillatory differential equation is

$$\ddot{x} + g(x) = \epsilon F(x, \dot{x}), \quad 0 < \epsilon \ll 1, \tag{7.8.8}$$

where both $g(x)$ and $F(x, \dot{x})$ are of odd parity, i.e.,

$$g(-x) = -g(x), \quad F(-x, -\dot{x}) = -F(x, \dot{x}), \tag{7.8.9}$$

and in almost all situations that arise in the natural and engineering sciences

$$F(x, \dot{x}) = \bar{F}(x^2, \dot{x}^2)\dot{x}. \tag{7.8.10}$$

- A very useful starting point is to study Eq. (7.8.8) using a first-order harmonic balance approximation. Such a calculation will allow us to determine the existence of limit-cycles and estimate the values of their amplitudes and angular frequencies.

- The Cveticanin method can now be used to calculate useful and hopefully accurate analytical approximations for the oscillatory solutions in the neighborhood of each limit-cycle.

Finally, it should be indicated that within the framework of the Cveticanin averaging method, information on the stability of the limit-cycles can be easily determined. Starting from the representation

$$x(t) = a(t, \epsilon) \cos \psi(t, \epsilon), \tag{7.8.11}$$

the averaging procedure gives a first-order differential equation for the function approximating $a(t, \epsilon)$,

$$\frac{da}{dt} = \epsilon H_1(a), \tag{7.8.12}$$

where $H_1(a)$ depends on the particular equation being studied and, in general,

$$H_1(0) = 0. \tag{7.8.13}$$

If limit-cycles exist, then their amplitudes correspond to the positive roots of the equation [1]

$$H_1(\bar{a}) = 0. \tag{7.8.14}$$

Assuming we have at least one root $\bar{a} > 0$, calculate

$$R(\bar{a}) \equiv \left. \frac{dH_1(a)}{da} \right|_{a=\bar{a}}.$$

The $R(\bar{a}) < 0$, the limit-cycle is locally stable; otherwise, it is unstable. (See Section 3.6 of Mickens [1] for the details of this analysis.)

7.9 Research Problems

We end this chapter and the book by presenting several outstanding research problems related to TNL oscillators.

(i) Professor Cveticanin [5] has constructed an appropriate generalization of the Krylov-Bogoliubov [6] method of averaging to terms of order ϵ. A further contribution can be made to this topic if the procedure could be extended to higher-order contributions in ϵ; see, for example [8].

(ii) For the equation

$$\ddot{x} + c_1 x|x|^{\alpha-1} = \epsilon F(x, \dot{x}),$$

assume that $F(x, \dot{x})$ can be written as

$$F(x, \dot{x}) = F_1(x, \dot{x})|\dot{x}|^{\beta}\text{sgn}(\dot{x}),$$

where

$$F_1(-x, -\dot{x}) = F_1(x, \dot{x}); \quad \beta > 0 \text{ and } \beta \neq 1.$$

The issue is whether the Cveticanin methodology can be applied to this situation? There may be few, if any, difficulties for an order ϵ calculation, but many problems may arise for higher-orders in ϵ.

(iii) The parameter-expansion and iteration methods appear to be related. What exactly is this connection and, if it exists, can it be used to improve results obtained by each technique?

(iv) Third-order harmonic balance generates three coupled, nonlinear algebraic equations which must be solved for two amplitudes and the angular frequency. In general, this is a very complex and difficult problem and the work to achieve this is computationally intensive. Can approximation techniques be created to resolve this problem?

(v) The ratio of neighboring coefficients of the harmonics in the periodic solutions (for the harmonic balance, parameter expansion and iteration methods) all decrease rapidly in the approximate solutions. In fact, the decrease is consistent with exponential decay. This strong decay of the coefficients appears to hold even for TNL oscillator equations for which this type of mathematical behavior is not *a priori* expected; an example is the equation

$$\ddot{x} + x^{1/3} = 0.$$

Question: Is this property of the coefficients an essential feature of these particular methods for determining approximations to the periodic solutions?

(vi) For the TNL oscillator equation

$$\ddot{x} + g(x) = 0,$$

does the "regularity," i.e., the number of derivatives for $g(x)$ existing at $x = 0$, affect the rate of decrease of the Fourier coefficients of the exact solution? If so, what is this relation? Also what impact does such a restriction have on the decay rate of the coefficients determined from any of the approximation methods?

References

[1] R. E. Mickens, *Nonlinear Oscillations* (Cambridge University Press, New York, 1991).
[2] R. E. Mickens, *Journal of Sound and Vibration* **259**, 457 (2003).
[3] R. E. Mickens and K. Oyedeji, *Journal of Sound and Vibration* **102**, 579 (1985).
[4] R. E. Mickens, *Journal of Sound and Vibration* **264**, 1195 (2003).
[5] L. Cveticanin, *Journal of Sound and Vibration* **320**, 1064 (2008).
[6] N. Krylov and N. Bogoliubov, *Introduction to Nonlinear Mechanics* (Princeton University Press; Princeton, NJ; 1943).
[7] R. E. Mickens, *Journal of Sound and Vibration* **292**, 964 (2006).
[8] N. Bogoliubov and J. A. Mitropolsky, *Asymptotical Methods in the Theory of Nonlinear Oscillations* (Hindustan Publishing Co.; Delhi, India; 1963).

Appendix A

Mathematical Relations

This appendix gives mathematical relations that are used regularly in the calculations of the text. The references listed at the end of this appendix contain extensive tables of other useful mathematical relations and analytic expressions.

A.1 Trigonometric Relations

A.1.1 *Exponential Definitions of Trigonometric Functions*

$$\sin x = \frac{e^{ix} - e^{-ix}}{2i} \tag{A.1}$$

$$\cos x = \frac{e^{ix} + e^{-ix}}{2}. \tag{A.2}$$

A.1.2 *Functions of Sums of Angles*

$$\sin(x \pm y) = \sin x \cos y \pm \cos x \sin y \tag{A.3}$$

$$\cos(x \pm y) = \cos x \cos y \mp \sin x \sin y. \tag{A.4}$$

A.1.3 *Powers of Trigonometric Functions*

$$\sin^2 x = \left(\frac{1}{2}\right)(1 - \cos 2x) \tag{A.5}$$

$$\cos^2 x = \left(\frac{1}{2}\right)(1 + \cos 2x) \tag{A.6}$$

$$\sin^3 x = \left(\frac{1}{4}\right)(3\sin x - \sin 3x) \tag{A.7}$$

$$\cos^3 x = \left(\frac{1}{4}\right)(3\cos x + \cos 3x) \tag{A.8}$$

$$\sin^4 x = \left(\frac{1}{8}\right)(3 - 4\cos 2x + \cos 4x) \tag{A.9}$$

$$\cos^4 x = \left(\frac{1}{8}\right)(3 + 4\cos 2x + \cos 4x) \tag{A.10}$$

$$\sin^5 x = \left(\frac{1}{16}\right)(10\sin x - 5\sin 3x + \sin 5x) \tag{A.11}$$

$$\cos^5 x = \left(\frac{1}{16}\right)(10\cos x + 5\cos 3x + \cos 5x) \tag{A.12}$$

$$\sin^6 x = \left(\frac{1}{32}\right)(10 - 15\cos 2x + 6\cos 4x - \cos 6x) \tag{A.13}$$

$$\cos^6 x = \left(\frac{1}{32}\right)(10 + 15\cos 2x + 6\cos 4x + \cos 6x). \tag{A.14}$$

A.1.4 *Other Trigonometric Relations*

$$\sin x \pm \sin y = 2\sin\left(\frac{x \pm y}{2}\right)\cos\left(\frac{x \mp y}{2}\right) \tag{A.15}$$

$$\cos x + \cos y = 2\cos\left(\frac{x + y}{2}\right)\cos\left(\frac{x - y}{2}\right) \tag{A.16}$$

$$\cos x - \cos y = -2\sin\left(\frac{x + y}{2}\right)\sin\left(\frac{x - y}{2}\right) \tag{A.17}$$

$$\sin x \cos y = \left(\frac{1}{2}\right)[\sin(x + y) + \sin(x - y)] \tag{A.18}$$

$$\cos x \sin y = \left(\frac{1}{2}\right)[\sin(x + y) - \sin(x - y)] \tag{A.19}$$

$$\cos x \cos y = \left(\frac{1}{2}\right)[\cos(x + y) + \cos(x - y)] \tag{A.20}$$

$$\sin x \sin y = \left(\frac{1}{2}\right)[\cos(x - y) - \cos(x + y)] \tag{A.21}$$

$$\sin^2 x - \sin^2 y = \sin(x + y)\sin(x - y) \tag{A.22}$$

$$\cos^2 x - \cos^2 y = -\sin(x + y)\sin(x - y) \tag{A.23}$$

$$\cos^2 x - \sin^2 y = \cos(x + y)\cos(x - y) \tag{A.24}$$

$$\sin^2 x \cos x = \left(\frac{1}{4}\right)(\cos x - \cos 3x) \tag{A.25}$$

$$\sin x \cos^2 x = \left(\frac{1}{4}\right)(\sin x + \sin 3x) \tag{A.26}$$

$$\sin^3 x \cos x = \left(\frac{1}{8}\right)(2\sin 2x - \sin 4x) \tag{A.27}$$

$$\sin^2 x \cos^2 x = \left(\frac{1}{8}\right)(1 - \cos 4x) \tag{A.28}$$

$$\sin x \cos^3 x = \left(\frac{1}{8}\right)(2\sin 2x + \sin 4x) \tag{A.29}$$

$$\sin^4 x \cos x = \left(\frac{1}{16}\right)(2\cos x - 3\cos 3x + \cos 5x) \tag{A.30}$$

$$\sin^3 x \cos^2 x = \left(\frac{1}{16}\right)(2\sin x + \sin 3x - \sin 5x) \tag{A.31}$$

$$\sin^2 x \cos^3 x = -\left(\frac{1}{16}\right)(2\cos x + \cos 3x + \cos 5x) \tag{A.32}$$

$$\sin x \cos^4 x = \left(\frac{1}{16}\right)(2\sin x + 3\sin 3x + \sin 5x). \tag{A.33}$$

A.1.5 *Derivatives and Integrals of Trigonometric Functions*

$$\frac{d}{dx}\cos x = -\sin x \tag{A.34}$$

$$\frac{d}{dx}\sin x = \cos x \tag{A.35}$$

$$\int \cos x \, dx = \sin x \tag{A.36}$$

$$\int \sin x \, dx = -\cos x \tag{A.37}$$

$$\int \sin^2 x \, dx = \left(\frac{1}{2}\right)x - \left(\frac{1}{4}\right)\sin 2x \tag{A.38}$$

$$\int \cos^2 x \, dx = \left(\frac{1}{2}\right)x + \left(\frac{1}{4}\right)\sin 2x \tag{A.39}$$

$$\int \sin mx \sin kx \, dx = \frac{\sin(m-k)x}{2(m-k)} - \frac{sin(m+k)x}{2(m+k)} \qquad m^2 \neq k^2 \quad \text{(A.40)}$$

$$\int \cos mx \cos kx \, dx = \frac{\sin(m-k)x}{2(m-k)} + \frac{sin(m+k)x}{2(m+k)} \qquad m^2 \neq k^2 \quad \text{(A.41)}$$

$$\int \sin mx \cos kx \, dx = -\frac{\cos(m-k)x}{2(m-k)} - \frac{cos(m+k)x}{2(m+k)} \qquad m^2 \neq k^2 \quad \text{(A.42)}$$

$$\int_{-\pi}^{\pi} \cos mx \cos kx \, dx = \pi \delta_{mk}; \qquad m, k \text{ integers} \quad \text{(A.43)}$$

$$\int_{-\pi}^{\pi} \sin mx \cos kx \, dx = 0; \qquad m, k \text{ integers} \quad \text{(A.44)}$$

$$\int_{-\pi}^{\pi} \sin mx \cos kx \, dx = 0; \qquad m, k \text{ integers} \quad \text{(A.45)}$$

$$\int x \sin x \, dx = \sin x - x \cos x \quad \text{(A.46)}$$

$$\int x^2 \sin x \, dx = 2x \sin x - (x^2 - 2) \cos x \quad \text{(A.47)}$$

$$\int x \cos x \, dx = \cos x + x \sin x \quad \text{(A.48)}$$

$$\int x^2 \cos x \, dx = 2x \cos x + (x^2 - 2) \sin x. \quad \text{(A.49)}$$

A.2 Factors and Expansions

$$(a \pm b)^2 = a^2 \pm 2ab + b^2 \quad \text{(A.50)}$$

$$(a \pm b)^3 = a^3 \pm 3a^2 b + 3ab^2 \pm b^3 \quad \text{(A.51)}$$

$$(a+b+c)^2 = a^2 + b^2 + c^2 + 2(ab + ac + bc) \quad \text{(A.52)}$$

$$(a+b+c)^3 = a^3 + b^3 + c^3 + 3a^2(b+c) + 3b^2(a+c) + 3c^2(a+b) + 6abc \quad \text{(A.53)}$$

$$a^2 - b^2 = (a-b)(a+b) \quad \text{(A.54)}$$

$$a^2 + b^2 = (a+ib)(a-ib), \qquad i = \sqrt{-1} \quad \text{(A.55)}$$

$$a^3 - b^3 = (a-b)(a^2 + ab + b^2) \quad \text{(A.56)}$$

$$a^3 + b^3 = (a+b)(a^2 - ab + b^2). \quad \text{(A.57)}$$

A.3 Quadratic Equations

The quadratic equation

$$ax^2 + bx + c = 0 \tag{A.58}$$

has the two solutions

$$x_1 = \frac{-b + \sqrt{b^2 - 4ac}}{2a}, \tag{A.59}$$

$$x_2 = \frac{-b - \sqrt{b^2 - 4ac}}{2a}. \tag{A.60}$$

A.4 Cubic Equations

The cube equation

$$z^3 + pz^2 + qz + r = 0 \tag{A.61}$$

can be reduced to the form

$$x^3 + ax + b = 0 \tag{A.62}$$

by substituting for z the value

$$z = x - \frac{p}{3}. \tag{A.63}$$

The constants a and b are given by the expressions

$$a = \frac{3q - p^2}{3}, \tag{A.64}$$

$$b = \frac{3p^3 - 9pq + 27r}{27}. \tag{A.65}$$

Let A and B be defined as

$$A = \left[-\left(\frac{b}{2}\right) + \left(\frac{b^2}{4} + \frac{a^3}{27}\right)^{1/2} \right]^{1/3}, \tag{A.66}$$

$$B = \left[-\left(\frac{b}{2}\right) - \left(\frac{b^2}{4} + \frac{a^3}{27}\right)^{1/2} \right]^{1/3}. \tag{A.67}$$

The three roots of Eq. (A.62) are given by the following expressions

$$x_1 = A + B, \tag{A.68}$$

$$x_2 = -\left(\frac{A+B}{2}\right) + \sqrt{-3}\left(\frac{A-B}{2}\right), \tag{A.69}$$

$$x_3 = -\left(\frac{A+B}{2}\right) - \sqrt{-3}\left(\frac{A-B}{2}\right). \tag{A.70}$$

Let

$$\Delta = \frac{b^2}{4} + \frac{a^3}{27}. \tag{A.71}$$

If $\Delta > 0$, then there will be one real root and two complex conjugate roots. If $\Delta = 0$, there will be three real roots, of which at least two are equal. If $\Delta < 0$, there will be three real and unequal roots.

A.5 Differentiation of a Definite Integral with Respect to a Parameter

Let $f(x,t)$ be continuous and have a continuous derivative $\partial f/\partial t$, in a domain in the x-t plane that includes the rectangle

$$\psi(t) \le x \le \phi(t), \qquad t_1 \le t \le t_2. \tag{A.72}$$

In addition, let $\psi(t)$ and $\phi(t)$ be defined and have continuous first derivatives for $t_1 \le t \le t_2$. Then, for $t_1 \le t \le t_2$, we have

$$\frac{d}{dt}\int_{\psi(t)}^{\phi(t)} f(x,t)dx = f[\phi(t),t]\frac{d\phi}{dt} - f[\psi(t),t]\frac{d\psi}{dt} + \int_{\psi(t)}^{\phi(t)} \frac{\partial}{\partial t} f(x,t)dx. \tag{A.73}$$

A.6 Eigenvalues of a 2 × 2 Matrix

The eigenvalues of a matrix A are given by the solutions to the characteristic equation

$$\det(A - \lambda I) = 0, \tag{A.74}$$

where I is the identity or unit matrix. If A is an $n \times n$ matrix, then there exists n eigenvalues λ_i, where $i = 1, 2, \ldots, n$.

Consider the 2×2 matrix

$$A = \begin{pmatrix} a & b \\ c & d \end{pmatrix}. \tag{A.75}$$

The characteristic equation is

$$\det\begin{pmatrix} a-\lambda & b \\ c & d-\lambda \end{pmatrix} = 0. \tag{A.76}$$

Evaluating the determinant gives

$$\lambda^2 - T\lambda + D = 0, \tag{A.77}$$

where

$$\begin{aligned} T &\equiv \text{trace}(A) = a + d, \\ D &\equiv \det(A) = ad - bc. \end{aligned} \tag{A.78}$$

The two eigenvalues are given by the expressions

$$\lambda_1 = \left(\frac{1}{2}\right)\left[T + \sqrt{T^2 - 4D}\right], \tag{A.79a}$$

$$\lambda_2 = \left(\frac{1}{2}\right)\left[T - \sqrt{T^2 - 4D}\right]. \tag{A.79b}$$

References

1. A. Erdélyi, *Tables of Integral Transforms, Vol. I* (McGraw-Hill, New York, 1954).
2. I. S. Gradshteyn and I. M. Ryzhik, *Table of Integrals, Series and Products* (Academic, New York, 1965).
3. E. Jaknke and F. Emde, *Tables of Functions with Formulas and Curves* (Dover, New York, 1943).
4. National Bureau of Standards, *Handbook of Mathematical Functions* (U.S. Government Printing Office; Washington, DC; 1964).
5. Chemical Rubber Company, *Standard Mathematical Tables* (Chemical Rubber Publishing Company, Cleveland, various editions).
6. H. B. Dwight, *Tables of Integrals and Other Mathematical Data* (MacMillan, New York, 1961).

Appendix B

Gamma and Beta Functions

B.1 Gamma Function

$$\Gamma(z) \equiv \int_0^\infty t^{z-1} e^{-t} dt, \quad \text{Re}(z) > 0 \tag{B.1}$$

$$\Gamma(z+1) = z\Gamma(z)m, \quad z \neq 0, -1, -2, -3, \ldots. \tag{B.2}$$

$$\Gamma(n+1) = n!, \quad n = 0, 1, 2, 3 \ldots. \tag{B.3}$$

$$\pi = 3.14\,159\,265\,358\,979$$

$$\frac{1}{\sqrt{2\pi}} = 0.39\,894\,228\,040\,143$$

$$e = 2.71\,828\,182\,845\,904$$

$$\Gamma\left(\frac{1}{2}\right) = \sqrt{\pi} = 1.77\,245\,358\,090\,551$$

$$\Gamma\left(\frac{1}{3}\right) = 2.67\,893\,85347$$

$$\Gamma\left(\frac{1}{4}\right) = 3.6256099082$$

B.2 The Beta Function

$$B(p,q) \equiv \int_0^1 t^{p-1}(1-t)^{q-1} dt \tag{B.4}$$

$$\text{Re}(p) > 0, \quad \text{Re}(q) > 0$$

$$B(p,q) = B(q,p) \tag{B.5}$$

$$B(p,q) = \frac{\Gamma(p)\Gamma(q)}{\Gamma(p+q)} \tag{B.6}$$

B.3 Two Useful Integrals

$$\int_0^1 \sqrt{1-t^p}\,dt = \left(\frac{1}{p}\right) B\left(\frac{3}{2}, \frac{1}{p}\right) \tag{B.7}$$

$$\begin{aligned}\int_0^{\pi/2} (\sin\theta)^m d\theta &= \int_0^{\pi/2} (\cos\theta)^m d\theta \\ &= \left(\frac{\sqrt{\pi}}{2}\right) \frac{\Gamma\left(\frac{m+1}{2}\right)}{\Gamma\left(\frac{m+2}{2}\right)} \\ &= \left(\frac{1}{2}\right) B\left(\frac{m+1}{2}, \frac{1}{2}\right)\end{aligned} \tag{B.8}$$

$$\int_0^{\pi/2} (\cos\theta)^{\nu-1} \cos(a\theta)\,d\theta = \frac{\pi}{(2^\nu)\nu B\left(\frac{\nu+a+1}{2}, \frac{\nu-a+1}{2}\right)} \tag{B.9}$$

Appendix C

Fourier Series

C.1 Definition of Fourier Series

Let $f(x)$ be a function that is defined on the interval $-L < x < L$ and is such that the following integrals exist:

$$\int_{-L}^{L} f(x)\cos\left(\frac{n\pi x}{L}\right)dx, \qquad \int_{-L}^{L} f(x)\sin\left(\frac{n\pi x}{L}\right)dx, \tag{C.1}$$

for $n = 0, 1, 2, \ldots$. The series

$$\frac{a_0}{2} + \sum_{n=1}^{\infty}\left[a_n\cos\left(\frac{n\pi x}{L}\right) + b_n\sin\left(\frac{n\pi x}{L}\right)\right] \tag{C.2}$$

where

$$a_n = \left(\frac{1}{L}\right)\int_{-L}^{L} f(x)\cos\left(\frac{n\pi x}{L}\right)dx, \tag{C.3}$$

$$b_n = \left(\frac{1}{L}\right)\int_{-L}^{L} f(x)\sin\left(\frac{n\pi x}{L}\right)dx, \tag{C.4}$$

is called the Fourier series of $f(x)$ on the interval $-L < x < L$. The numbers $\{a_n\}$ and $\{b_n\}$ are called the Fourier coefficients of $f(x)$.

A function $f_1(x)$ such that

$$f_1(x+p) = f_1(x), \qquad p \neq 0, \tag{C.5}$$

for all x is said to be periodic and to have period p.

Since both $\sin(n\pi x/L)$ and $\cos(n\pi x/L)$ have period $2L/n$, the only period shared by all these expressions is $2L$. Therefore, if the Fourier series of $f(x)$ converges, then $f(x)$ is periodic of period $2L$, i.e.,

$$f(x+2L) = f(x). \tag{C.6}$$

If $f(x)$ is initially defined only in the interval $-L < x < L$, then Eq. (C.6) can be used to define it for all values of x, i.e., $-\infty < x < \infty$.

In general, the Fourier series of $f(x)$ defined on an interval $-L < x < L$ is a strictly formal expansion. The next section gives the relevant theorem concerning convergence of Fourier series.

C.2 Convergence of Fourier Series

A function $f(x)$ is said to be piecewise smooth on a finite interval $a \leq x \leq b$ if this interval can be divided into a finite number of subintervals such that (1) $f(x)$ has a continuous derivative in the interior of each of these subintervals, and (2) both $f(x)$ and df/dx approach finite limits as x approaches either end point of each of these subintervals from its interior.

C.2.1 *Examples*

The function $f(x)$ defined by

$$f(x) = \begin{cases} \pi, & -\pi \leq x < 0, \\ x, & 0 < x \leq \pi, \end{cases} \tag{C.7}$$

is piecewise smooth on the interval $-\pi < x < \pi$. The two subintervals are $[-\pi, 0)$ and $(0, \pi]$.

The function $f(x)$ defined on the interval $0 \leq x \leq 5$ by

$$f(x) = \begin{cases} x^2, & 0 \leq x < 1, \\ 2 - x, & 1 \leq x < 3, \\ 1, & 3 \leq x < 4, \\ (x-4)^{3/2}, & 4 < x \leq 5, \end{cases} \tag{C.8}$$

is piecewise smooth on this interval. Observe that in each subinterval both $f(x)$ and df/dx are defined.

C.2.2 *Convergence Theorem*

Theorem C.1. *Let $f(x)$, (1) be periodic of period $2L$, and (2) be piecewise smooth on the interval $-L < x < L$. Then the Fourier series of $f(x)$*

$$\frac{a_0}{2} + \sum_{n=1}^{\infty} \left[a_n \cos\left(\frac{n\pi x}{L}\right) + b_n \sin\left(\frac{b\pi x}{L}\right) \right], \tag{C.9}$$

where

$$a_n = \left(\frac{1}{L}\right)\int_{-L}^{L} f(x)\cos\left(\frac{n\pi x}{L}\right)dx, \tag{C.10}$$

$$b_n = \left(\frac{1}{L}\right)\int_{-L}^{L} f(x)\sin\left(\frac{n\pi x}{L}\right)dx, \tag{C.11}$$

converges at every point x_0 *to the value*

$$\frac{f(x_0^+) + f(x_0^-)}{2}, \tag{C.12}$$

where $f(x_0^+)$ *is the right-hand limit of* $f(x)$ *at* x_0 *and* $f(x_0^-)$ *is the left-hand limit of* $f(x)$ *at* x_0. *If* $f(x_0)$ *is continuous at* x_0, *the value given by Eq.* (C.12) *reduces to* $f(x_0)$ *and the Fourier series of* $f(x)$ *converges to* $f(x_0)$.

C.3 Bounds on Fourier Coefficients [1, 2, 7]

Theorem C.2. *Let* $f(x)$ *be periodic of period* $2L$ *and be piecewise smooth on the interval* $-L < x < L$. *Let the first* r *derivatives of* $f(x)$ *exist and let* $f(x)$ *be of bounded variation. Then there exists a positive constant* M *(whose value may depend on* $f(x)$ *and* L*) such that the Fourier coefficients satisfy the relation*

$$|a_n| + |b_n| \leq \frac{M}{n^{r+1}}. \tag{C.13}$$

Comments. A function $f(x)$, defined on $-L < x < L$, is of bounded variation if the arc-length of $f(x)$ over this interval is bounded [2].

Theorem C.3. *Let* $f(x)$ *be analytic in* x *and be periodic with period* $2L$. *There exist a* θ *and an* A *(which may depend on* $f(x)$ *and* $2L$*) such that the Fourier coefficients satisfy the relation*

$$|a_n| + |b_n| \leq A\theta^n, \quad 0 < \theta < 1. \tag{C.14}$$

C.4 Expansion of $F(a\cos x, -a\sin x)$ in a Fourier Series

At a number of places in the text, the Fourier series is needed for a function of two variables, $F(u, v)$, where

$$u = a\cos x, \qquad v = -a\sin x, \tag{C.15}$$

and in general $F(u,v)$ is a polynomial function of u and v. To illustrate how this is done, consider the following particular form for $F(u,v)$:

$$F(u,v) = (1-u^2)v. \tag{C.16}$$

Replacing u and v by the relations of Eq. (C.15), and using the trigonometric relations given in Appendix A, the following result is obtained:

$$\begin{aligned} F(u,v) &= (1-u^2)v = (1-a^2\cos^2 x)(-a\sin x) = -a\sin x + a^3\cos^2 x\sin x \\ &= -a\sin x + \left(\frac{a^3}{4}\right)(\sin x + \sin 3x) \\ &= \left(\frac{a^2-4}{4}\right)a\sin x + \left(\frac{a^3}{4}\right)\sin 3x. \end{aligned} \tag{C.17}$$

This last expression is the required Fourier expansion of Eq. (C.16).

For a second example, consider $F(u,v) = u^3$. The following is obtained for this case:

$$F(u,v) = u^3 = a^3\cos^3 x = \left(\frac{3a^3}{4}\right)\cos x + \left(\frac{a^3}{4}\right)\cos 3x. \tag{C.18}$$

C.5 Fourier Series for $(\cos\theta)^\alpha$ and $(\sin\theta)^\alpha$

See Appendix B, Eq. (B.9) for a useful integral relation that can be used to derive the following relations:

Let a_{2p+1} be defined as

$$a_{2p+1} = \frac{3\Gamma\left(\frac{7}{3}\right)}{\left(2^{4/3}\right)\Gamma\left(p+\frac{5}{3}\right)\Gamma\left(\frac{2}{3}-p\right)}, \tag{C.19}$$

for $p = 0, 1, 2, \ldots$, with

$$a_1 = \frac{\Gamma\left(\frac{1}{3}\right)}{2^{1/3}\left[\Gamma\left(\frac{2}{3}\right)\right]^2} = 1.159595266963929. \tag{C.20}$$

Then $(\cos\theta)^{1/3}$ and $(\sin\theta)^{1/3}$ have the following Fourier series

$$\begin{aligned} (\cos\theta)^{1/3} &= \sum_{p=0}^{\infty} a_{2p+1}\cos(2p+1)\theta = a_1\left[\cos\theta - \frac{\cos(3\theta)}{5} + \frac{\cos(5\theta)}{10}\right. \\ &\quad - \frac{7\cos(7\theta)}{110} + \frac{\cos(9\theta)}{22} - \frac{13\cos(11\theta)}{374} + \frac{26\cos(13\theta)}{935} \\ &\quad \left. - \frac{494\cos(15\theta)}{21505} + \cdots\right] \end{aligned} \tag{C.21}$$

$$(\sin\theta)^{1/3} = \sum_{p=0}^{\infty}(-1)^p a_{2p+1}\sin(2p+1)\theta = a_1\left[\sin\theta + \frac{\sin(3\theta)}{5} + \frac{\sin(5\theta)}{10} + \frac{7\sin(7\theta)}{110} + \frac{\sin(9\theta)}{22} + \frac{13\sin(11\theta)}{374} + \frac{26\sin(13\theta)}{935} + \frac{494\sin(15\theta)}{21505} + \cdots\right] \quad (C.22)$$

Similarly, we have

$$(\cos\theta)^{2/3} = a_0\left[\frac{1}{2} + \frac{\cos(2\theta)}{4} - \frac{\cos(4\theta)}{14} + \frac{\cos(6\theta)}{28} - \frac{2\cos(8\theta)}{91} + \frac{11\cos(10\theta)}{728} - \frac{11\cos(12\theta)}{988} + \frac{17\cos(14\theta)}{1976} - \frac{17\sin(16\theta)}{2470} + \cdots\right], \quad (C.23)$$

$$(\sin\theta)^{2/3} = a_0\left[\frac{1}{2} + \frac{\cos(2\theta)}{4} - \frac{\cos(4\theta)}{14} + \frac{\cos(6\theta)}{28} - \frac{2\cos(8\theta)}{91} + \frac{11\cos(10\theta)}{728} - \frac{11\cos(12\theta)}{988} + \frac{17\cos(14\theta)}{1976} - \frac{17\sin(16\theta)}{2470} + \cdots\right], \quad (C.24)$$

where

$$a_0 = \frac{3\cdot 2^{4/3}\Gamma\left(\frac{2}{3}\right)}{\left[\Gamma\left(\frac{1}{3}\right)\right]^2} = 1.426348256. \quad (C.25)$$

Another useful Fourier expansion relation is

$$|\cos\theta| = \left(\frac{4}{\pi}\right)\left[\frac{1}{2} + \frac{\cos(2\theta)}{3} - \frac{\cos(4\theta)}{15} + \cdots\right]$$

as well as

$$[\mathrm{sign}(\cos\theta)]|\cos\theta|^{\alpha} = a_1\cos\theta + a_2\cos(3\theta) + a_5\cos(5\theta) + \cdots, \quad (C.26)$$

where

$$\begin{cases} a_1 = \dfrac{4\Gamma\left(1+\frac{\alpha}{2}\right)}{\sqrt{\pi}(\alpha+1)\Gamma\left(\frac{\alpha+1}{2}\right)}, \\ a_3 = \left(\dfrac{\alpha-1}{\alpha+3}\right)a_1, \\ a_5 = \left[\dfrac{(\alpha-1)(\alpha-3)}{(\alpha+3)(\alpha+5)}\right]a_1, \end{cases} \quad (C.27)$$

and

$$\frac{1}{\cos\theta} = 2\sum_{p=0}^{\infty}(-1)^p\cos(2p+1)\theta. \quad (C.28)$$

References

1. N. K. Bary, *A Treatise on Trigonometric Series*, Vol. I (MacMillan, New York, 1964).
2. R. C. Buck, *Advanced Calculus* (McGraw-Hill, New York, 1978).
3. H. S. Carslaw, *Theory of Fourier Series and Integrals* (MacMillan, London, 1921).
4. R. V. Churchill, *Fourier Series and Boundary Value Problems* (McGraw-Hill, New York, 1941).
5. W. Kaplan, *Advanced Calculus* (Addison-Wesley; Reading, MA; 1952). See Chapter 7.
6. W. Rogosinski, *Fourier Series* (Chelsea Publishing, New York, 1950).
7. E. C. Titchmarch, *Eigenfunction Expansions* (Oxford University Press, Oxford, 1946).
8. A. Zygmund, *Trigonometrical Series* (Dover, New York, 1955).

Appendix D

Basic Theorems of the Theory of Second-Order Differential Equations

D.1 Introduction

The general second-order differential equation

$$\frac{d^2y}{dt^2} = f\left(y, \frac{dy}{dt}, t\right), \tag{D.1}$$

can be written in the system form

$$\frac{dy_1}{dt} = y_2, \tag{D.2}$$

$$\frac{dy_2}{dt} = f(y_1, y_2, t), \tag{D.3}$$

by means of the transformation $(y, dy/dt) = (y_1, y_2)$. A general system of coupled, first-order differential equations is

$$\frac{dy_1}{dt} = f_1(y_1, y_2, t), \tag{D.4}$$

$$\frac{dy_2}{dt} = f_2(y_1, y_2, t). \tag{D.5}$$

In this appendix, a number of theorems are given concerning the solutions of Eqs. (D.4) and (D.5). Proofs can be found in the references listed at the end of this appendix.

The following assumptions and definitions apply to all the results of this appendix: (1) The functions $f_1(y_1, y_2, t)$ and $f_2(y_1, y_2, t)$, defined in a certain domain R of the three-dimensional (y_1, y_2, t) space, are continuous in this region, and have continuous partial derivatives with respect to y_1, y_2 and t. (2) A point having the coordinates $(\bar{y}_1, \bar{y}_2, \bar{t})$ will be denoted as $P(\bar{y}_1, \bar{y}_2, \bar{t})$.

D.2 Existence and Uniqueness of the Solution

Theorem D.1. *Let $P(y_1^0, y_2^0, t_0)$ be any point in R. There exists an interval of t $(t_1 < t < t_2)$ containing t_0, and only one set of functions*

$$y_1 = \phi_1(t), \qquad y_2 = \phi_2(t), \tag{D.6}$$

defined in this interval, for which the following conditions are satisfied: (1) $\phi_1(t_0) = y_1^0$ and $\phi_2(t_0) = y_2^0$. (2) For all values of t in the interval, $t_1 < t < t_2$, the point $P[\phi_1(t), \phi_2(t), t]$ belongs to the domain R. (3) The system of functions given by Eq. (D.6) satisfies the system of differential equations Eq. (D.4) and Eq. (D.5). (4) The solutions, given by Eq. (D.6), can be continued up to the boundary of the domain R; that is, whatever closed domain $\bar{R}_1$, contained entirely in R, there are values t' and t'', where

$$t_1 < t' < t_2, \qquad t_1 < t'' < t_2, \tag{D.7}$$

such that the points $P[\phi_1(t'), \phi_2(t'), t']$ and $P[\phi_1(t''), \phi_2(t''), t'']$ lie outside $\bar{R}_1$.

D.3 Dependence of the Solution on Initial Conditions

The solutions of Eqs. (D.4) and (D.5) depend on the initial conditions (y_1^0, y_2^0, t_0). Consequently, the solutions can be written as

$$y_1 = \phi_1(t, t_0, y_1^0, y_2^0), \qquad y_2 = \phi_2(t, t_0, y_1^0, y_2^0), \tag{D.8}$$

with

$$y_1^0 = \phi_1(t_0, t_0, y_1^0, y_2^0), \qquad y_2^0 = \phi_2(t_0, t_0, y_1^0, y_2^0). \tag{D.9}$$

The following theorems give information concerning the dependence of the solutions on the initial conditions.

Theorem D.2. *Let*

$$y_1 = \phi_1(t, t^*, y_1^*, y_2^*), \tag{D.10a}$$
$$y_2 = \phi_2(t, t^*, y_1^*, y_2^*), \tag{D.10b}$$

be a solution to Eqs. (D.4) and (D.5), defined for t in the interval, $t_1 < t < t_2$, and having the initial values

$$y_1(t^*) = y_1^*, \qquad y_2(t^*) = y_2^*. \tag{D.11}$$

Let T_1 and T_2 be arbitrary numbers satisfying the condition

$$t_1 < T_1 < T_2 < t_2. \tag{D.12}$$

Then for an arbitrary positive ϵ, there exists a positive number $\delta = \delta(\epsilon, T_1, T_2)$ such that for the values of t_0, y_1^0 and y_2^0 for which

$$|t_0 - t^*| < \delta, \qquad |y_1^0 - y_1^*| < \delta, \qquad |y_2^0 - y_2^*| < \delta, \tag{D.13}$$

the solutions

$$y_1 = \phi_1(t, t_0, y_1^0, y_2^0), \qquad y_2 = \phi_2(t, t_0, y_1^0, y_2^0), \tag{D.14}$$

are defined for all values of t in the interval $T_1 \leq t \leq T_2$, and satisfy the inequalities

$$|\phi_1(t, t_0, y_1^0, y_2^0) - \phi_1(t, t^*, y_1^*, y_2^*)| < \epsilon, \tag{D.15}$$

$$|\phi_2(t, t_0, y_1^0, y_2^0) - \phi_2(t, t^*, y_1^*, y_2^*)| < \epsilon. \tag{D.16}$$

Theorem D.3. *If the functions $f_1(y_1, y_2, t)$ and $f_2(y_1, y_2, t)$ of Eqs.* (D.4) *and* (D.5) *have continuous partial derivatives with respect to the variables y_1 and y_2 of order up to $n \geq 1$, then the solutions to this system have continuous partial derivatives with respect to y_1^0 and y_2^0 of the same order.*

Theorem D.4. *If the functions $f_1(y_1, y_2, t)$ and $f_2(y_1, y_2, t)$ are analytic functions of the variables y_1 and y_2, then the solution, given by Eq.* (D.8), *is an analytic function of its arguments in a neighborhood of every set of values for which the functions $f_1(y_1, y_2, t)$ and $f_2(y_1, y_2, t)$ are defined.*

D.4 Dependence of the Solution on a Parameter

Let the functions f_1 and f_2 depend on a parameter λ. For this case, Eqs. (D.4) and (D.5) become

$$\frac{dy_1}{dt} = f_1(y_1, y_2, t, \lambda), \tag{D.17}$$

$$\frac{dy_2}{dt} = f_2(y_1, y_2, t, \lambda). \tag{D.18}$$

Theorem D.5. *If the functions $f_1(y_1, y_2, t, \lambda)$ and $f_2(y_1, y_2, t, \lambda)$ are continuous functions of λ, the solutions of Eqs.* (D.17) *and* (D.18)

$$y_1 = \phi_1(t, t_0, y_1^0, y_2^0, \lambda), \qquad y_2 = \phi_2(t, t_0, y_1^0, y_2^0, \lambda), \tag{D.19}$$

are also continuous functions of λ.

Theorem D.6. *Let* $f_1(y_1, y_2, t, \lambda)$ *and* $f_2(y_1, y_2, t, \lambda)$, *and the first partial derivatives of* f_1 *and* f_2, *with respect to* y_1 *and* y_2, *be continuous functions of* λ. *If* y_1 *and* y_2, *given by Eq.* (D.19) *are solutions of Eqs.* (D.17) *and* (D.18), *then the derivatives*

$$\frac{\partial \phi_i(t, t_0, y_1^0, y_2^0, \lambda)}{\partial y_j^0}, \quad i = (1,2), \quad j = (1,2), \tag{D.20}$$

are also continuous functions of λ.

Theorem D.7. *If* $f_1(y_1, y_2, t, \lambda)$ *and* $f_2(y_1, y_2, t, \lambda)$ *are analytic functions of their arguments, then the solutions to Eqs.* (D.17) *and* (D.18) *are also analytic functions of all their arguments in a neighborhood of every set of values* $(t, t_0, y_1^0, y_2^0, \lambda)$ *for which they are defined.*

References

1. A. A. Andronov, A. A. Vitt and S. E. Khaikin, *Theory of Oscillators* (Addison-Wesley; Reading, MA; 1966). See the Appendix, pp. 795–800.
2. E. A. Coddington and N. Levinson, *Theory of Ordinary Differential Equations* (McGraw-Hill, New York, 1995). See Chapter 2.
3. E. L. Ince, *Ordinary Differential Equations* (Dover, New York, 1956).
4. N. Minorsky, *Nonlinear Oscillations* (Robert E. Krieger; Huntington, NY; 1962). See pp. 228–231.
5. S. L. Ross, *Differential Equations* (Blaisdell; Waltham, MA; 1964). See Chapters 10 and 11.
6. G. Sansone and R. Conti, *Nonlinear Differential Equations* (Pergamon, New York, 1964). Chapters VI and VII give excellent discussions of the topics presented in this appendix.

Appendix E

Linear Second-Order Differential Equations

Essentially all of the approximation methods given in this book eventually lead to linear, second-order differential equations. This appendix gives the basic theorems and rules for solving this type of differential equation. Detailed proofs of the various theorems can be found in the references given at the end of this appendix.

E.1 Basic Existence Theorem

The general linear, second-order differential equation takes the form

$$a_0(t)\,\frac{d^2y}{dt^2} + a_1(t)\frac{dy}{dt} + a_2(t)y = F(t). \tag{E.1}$$

If $F(t) = 0$, then Eq. (E.1) is said to be homogeneous; if $F(t) \neq 0$, then Eq. (E.1) is said to be inhomogeneous.

Theorem E.1. *Let $a_0(t)$, $a_1(t)$, $a_2(t)$ and $F(t)$ be continuous on the interval $a \leq t \leq b$, with $a_0(t) \neq 0$ on this interval. Let t_0 be a point of the interval $a \leq t \leq b$, and let C_1 and C_2 be two real constants. Then there exists a unique solution $y = \phi(t)$ of Eq.* (E.1) *such that*

$$\phi(t_0) = C_1, \qquad \frac{d\phi(t_0)}{dt} = C_2, \tag{E.2}$$

and the solution is defined over the entire interval $a \leq t \leq b$.

E.2 Homogeneous Linear Differential Equations

The linear second-order homogeneous differential equation has the form

$$a_0(t)\,\frac{d^2y}{dt^2} + a_1(t)\frac{dy}{dt} + a_2(t)y = 0. \tag{E.3}$$

Again, it is assumed that $a_0(t)$, $a_1(t)$, and $a_2(t)$ are continuous on the interval $a \leq t \leq b$ and $a_0(t) \neq 0$ on this interval.

Theorem E.2. *Let $\phi(t)$ be a solution of Eq.* (E.3) *such that*

$$\phi(t_0) = 0, \qquad \frac{d\phi(t_0)}{dt} = 0, \tag{E.4}$$

where $a \leq t_0 \leq b$. Then $\phi(t) = 0$ for all t in this interval.

To proceed the concepts of linear combination, linear dependence, and linear independence must be introduced and defined.

E.2.1 *Linear Combination*

If $f_1(t), f_2(t), \ldots, f_n(t)$ are n functions and $C_1, C_2, \ldots, C_n$ are n arbitrary constants, then the expression

$$C_1 f_1(t) + C_2 f_2(t) + \cdots + C_n f_n(t) \tag{E.5}$$

is called a linear combination of $f_1(t), f_2(t), \ldots, f_n(t)$.

E.2.2 *Linear Dependent and Linear Independent Functions*

The n functions $f_1(t), f_2(t), \ldots, f_n(t)$ are linearly dependent on $a \leq t \leq b$ if and only if there exist constants $C_1, C_2, \ldots, C_n$, not all zero, such that

$$C_1 f_1(t) + C_2 f_2(t) + \cdots + C_n f_n(t) = 0 \tag{E.6}$$

for all t such that $a \leq t \leq b$.

The n functions $f_1(t), f_2(t), \ldots, f_n(t)$ are linearly independent on $a \leq t \leq b$ if and only if they are not linearly dependent there; that is, $f_1(t), f_2(t), \ldots, f_n(t)$ are linearly independent on $a \leq t \leq b$ if and only if

$$C_1 f_1(t) + C_2 f_2(t) + \cdots + C_n f_n(t) = 0 \tag{E.7}$$

for all t such that $a \leq t \leq b$ implies that

$$C_1 = C_2 = \cdots = C_n = 0. \tag{E.8}$$

E.2.3 *Theorems on Linear Second-Order Homogeneous Differential Equations*

Theorem E.3. *Let the functions $f_1(t), f_2(t), \ldots, f_n(t)$ be any n solution of Eq.* (E.3) *on the interval $a \leq t \leq b$. Then the function*

$$C_1 f_1(t) + C_2 f_2(t) + \cdots + C_n f_n(t) \tag{E.9}$$

where $C_1, C_2, \ldots, C_n$ *are arbitrary constants, is also a solution of Eq.* (E.3) *on* $a \leq t \leq b$.

Theorem E.4. *There exists a set of two solutions of Eq.* (E.3) *such that the two solutions are linearly independent on* $a \leq t \leq b$.

Definition E.1. Let $f_1(t)$ and $f_2(t)$ be real functions, each of which has a derivative on $a \leq t \leq b$. The determinant

$$\begin{vmatrix} f_1(t) & f_2(t) \\ \dfrac{df_1(t)}{dt} & \dfrac{df_2(t)}{dt} \end{vmatrix} \tag{E.10}$$

is called the Wronskian of the two functions $f_1(t)$ and $f_2(t)$. Denote it by $W(f_1, f_2, t) \equiv W(t)$.

Theorem E.5. *Let* $f_1(t)$ *and* $f_2(t)$ *be two solutions of Eq.* (E.3) *on* $a \leq t \leq b$. *Let* $W(t)$ *denote the Wronskian of* $f_1(t)$ *and* $f_2(t)$. *Then either* $W(t)$ *is zero for all* t *on* $a \leq t \leq b$ *or* $W(t)$ *is zero for no* t *on* $a \leq t \leq b$. *The Wronskian* $W(t)$ *is zero if and only if the two solutions* $f_1(t)$ *and* $f_2(t)$ *are linearly dependent on* $a \leq t \leq b$.

Theorem E.6. *Let* $f_1(t)$ *and* $f_2(t)$ *be two linearly independent solutions of Eq.* (E.3) *on* $a \leq t \leq b$. *Let* $W(t)$ *be their Wronskian and let* $a \leq t \leq b$. *Then*

$$W(t) = W(t_0) \exp\left[-\int_{t_0}^{t} \frac{a_1(z)}{a_0(z)}\, dz\right] \tag{E.11}$$

for all t *on* $a \leq t \leq b$.

Theorem E.7. *Let* $f_1(t)$ *and* $f_2(t)$ *be any two linearly independent solutions of Eq.* (E.3) *on* $a \leq t \leq b$. *Every solution* $f(t)$ *of Eq.* (E.3) *can be expressed as a suitable linear combination of* $f_1(t)$ *and* $f_2(t)$, *i.e.*,

$$f(t) = C_1 f_1(t) + C_2 f_2(t), \tag{E.12}$$

where C_1 *and* C_2 *are arbitrary constants.*

E.3 Inhomogeneous Linear Differential Equations

The general linear, second-order, inhomogeneous differential equation takes the form

$$a_0(t)\,\frac{d^2y}{dt^2} + a_1(t)\,\frac{dy}{dt} + a_2(t)y = F(t). \tag{E.13}$$

It is assumed that $a_0(t)$, $a_1(t)$, $a_2(t)$ and $F(t)$ are continuous on $a \leq t \leq b$, with $a_0(t) \neq 0$ on this interval.

Equation (E.13) can be written as

$$Ly = F(t), \tag{E.14}$$

where L is the linear operator

$$L \equiv a_0(t)\frac{d^2}{dt^2} + a_1(t)\frac{d}{dt} + a_2(t). \tag{E.15}$$

Theorem E.8. *Let $v(t)$ be any solution of the inhomogeneous Eq.* (E.13), *and let $u(t)$ be any solution of the homogeneous equation*

$$Ly = 0. \tag{E.16}$$

Then $u(t) + v(t)$ is also a solution of the inhomogeneous Eq. (E.13).

The solution $u(t)$ is called the homogeneous part of the solution to Eq. (E.13), and $v(t)$ is called the particular solution to Eq. (E.13). The homogeneous solution $u(t)$ will contain two arbitrary constants. However, the particular solution $v(t)$ will not contain any arbitrary constants.

E.3.1 *Principle of Superposition*

The principle of superposition for linear second-order inhomogeneous differential equations is given in the following theorem.

Theorem E.9. *Let*

$$Ly = F_i(t), \qquad i = 1, 2, \ldots, n, \tag{E.17}$$

be n different inhomogeneous second-order differential equations where the linear operator L is defined by Eq. (E.15). *Let $f_i(t)$ be a particular solution of Eq.* (E.17) *for $i = 1, 2, \ldots, n$. Then*

$$\sum_{i=1}^{n} f_i(t) \tag{E.18}$$

is a particular solution of the equation

$$Ly = \sum_{i=1}^{n} F_i(t). \tag{E.19}$$

E.3.2 *Solutions of Linear Inhomogeneous Differential Equations*

Write Eq. (E.13) in "normal" form, i.e.,

$$\frac{d^2y}{dt^2} + p(t)\frac{dy}{dt} + q(t)y = f(t), \tag{E.20}$$

where $p(t)$, $q(t)$, and $f(t)$ are continuous functions for $a \le t \le b$. Assume that two linearly independent solutions, $y_1(t)$ and $y_2(t)$ are known for the corresponding homogeneous differential equation

$$\frac{d^2y}{dt^2} + p(t)\frac{dy}{dt} + q(t)y = 0. \tag{E.21}$$

The general solution of Eq. (E.20) is

$$y(t) = C_1y_1(t) + C_2y_2(t) + \frac{1}{W(t_0)}\int_{t_0}^{t} f(x)e^{I(x)}[y_1(x)y_2(t) - y_1(t)y_2(x)]dx, \tag{E.22}$$

where $W(t_0) = W(y_1, y_2, t_0)$ is the Wronskian of $y_1(t)$ and $y_2(t)$ evaluated at $t = t_0$, $a \le t \le b$; C_1 and C_2 are arbitrary constants; and $I(x)$ is

$$I(x) = \int_{x_0}^{x} p(z)dz. \tag{E.23}$$

E.4 Linear Second-Order Homogeneous Differential Equations with Constant Coefficients

For the special case of constant coefficients, the problem of obtaining two linear independent solutions of a homogeneous second-order differential equation can be completely solved.

Consider the differential equation

$$a_0\frac{d^2y}{dt^2} + a_1\frac{dy}{dt} + a_2y = 0, \tag{E.24}$$

where the coefficients a_0, a_1, and a_2 are real constants. The equation

$$a_0m^2 + a_1m + a_2 = 0 \tag{E.25}$$

is called the characteristic equation corresponding to Eq. (E.24). The two roots of Eq. (E.25), m_1 and m_2 are related to the general solution of Eq. (E.24) as follows:

(1) Let m_1 and m_2 be real and distinct, i.e., $m_1 \ne m_2$. The general solution of Eq. (E.24), in this case, is

$$y(t) = C_1e^{m_1t} + C_2e^{m_2t}, \tag{E.26}$$

where C_1 and C_2 are arbitrary constants.

(2) Let m_1 and m_2 be complex conjugates of each other, i.e., $m_1 = m_2^* = a + ib$. For this case, the general solution of Eq. (E.24) is either one of the following equivalent forms:

$$y(t) = \begin{cases} Ae^{at}\cos(bt + B), \\ C_1 e^{at}\cos bt + C_2 e^{at}\sin bt, \end{cases} \tag{E.27}$$

where A, B, C_1 and C_2 are arbitrary constants.

(3) Let m_1 and m_2 be equal, i.e., $m_1 = m_2 = m$. The general solution for this case is

$$y(t) = (C_1 + C_2 t)e^{mt}, \tag{E.28}$$

where C_1 and C_2 are arbitrary constants.

E.5 Linear Second-Order Inhomogeneous Differential Equations with Constant Coefficients

Consider the following inhomogeneous differential equation

$$a_0 \frac{d^2y}{dt^2} + a_1 \frac{dy}{dt} + a_2 y = Q(t), \tag{E.29}$$

where a_0, a_1, and a_2 are constants and $Q(t)$ has first and second derivatives for an interval $a \le t \le b$. In general, if $Q(t)$ takes the form of a sum of terms, each having the structure

$$Q_{n,k}(t) = t^n e^{kt}, \tag{E.30}$$

then the general solution to Eq. (E.29) can be found, i.e.,

$$y(t) = C_1 e^{m_1 t} + C_2 e^{m_2 t} + v(t) \tag{E.31}$$

where the homogeneous solution is

$$u(t) = C_1 e^{m_1 t} + C_2 e^{m_2 t} \tag{E.32}$$

and $v(t)$ is a solution to the inhomogeneous Eq. (E.29).

For the applications in this book, the following two rules will allow the determination of particular solution $v(x)$ to Eqs. (E.29) and (E.30).

Rule 1. Let no term of $Q(t)$ be the same as a term in the homogeneous solution $u(t)$. In this case, a particular solution of Eq. (E.29) will be a linear combination of the terms in $Q(t)$ and all its linearly independent derivatives.

The following example illustrates this rule. Consider the equation

$$\frac{d^2y}{dt^2} - 3\frac{dy}{dt} + 2y = 2te^{3t} + 3\sin t. \tag{E.33}$$

The characteristic equation is

$$m^2 - 3m + 2 = 0, \tag{E.34}$$

and has solutions $m_1 = 1$ and $m_2 = 2$. Therefore, the solution to the homogeneous equation is

$$u(t) = C_1e^t + C_2e^{2t}, \tag{E.35}$$

where C_1 and C_2 are arbitrary constants. Observe that no term of

$$Q(t) = 2te^{3t} + 3\sin t \tag{E.36}$$

is a member of the homogeneous solution. A particular solution of Eq. (E.33) will be a linear combination of $t\exp(3t)$ and $\sin t$, and their linearly independent derivatives $\exp(3t)$ and $\cos t$. Consequently, the particular solution $v(t)$ has the form

$$v(t) = Ate^{3t} + Be^{3t} + C\sin t + D\cos t, \tag{E.37}$$

where A, B, C, and D are constants. These constants can be determined by substituting Eq. (E.37) into Eq. (E.33) and setting the coefficients of the linearly independent terms, $t\exp(3t)$, $\exp(3t)$, $\sin t$, and $\cos t$, equal to zero. Doing this gives

$$A = 1, \quad B = -\left(\frac{3}{2}\right), \qquad C = \frac{3}{10}, \qquad D = \frac{9}{10}. \tag{E.38}$$

The particular solution $v(t)$ is

$$v(t) = te^{3t} - \left(\frac{3}{2}\right)e^{3t} + \left(\frac{3}{10}\right)\sin t + \left(\frac{9}{10}\right)\cos t, \tag{E.39}$$

and the general solution to Eq. (E.33) is

$$y(t) = C_1e^t + C_2e^{2t} + te^{3t} - \left(\frac{3}{2}\right)e^{3t} + \left(\frac{3}{10}\right)\sin t + \left(\frac{9}{10}\right)\cos t. \tag{E.40}$$

Rule 2. Let $Q(t)$, in Eq. (E.29), contain a term that, ignoring constant coefficients, is t^k times a term $u_1(t)$ of $u(t)$, where k is zero or a positive integer. In this case, a particular solution to Eq. (E.29) will be a linear combination of $t^{k+1}u_1(t)$ and all its linearly independent derivatives that are *not* contained in $u(t)$.

As an illustration of this rule, consider the equation

$$\frac{d^2y}{dt^2} - 3\frac{dy}{dt} + 2y = 2t^2 + 3e^{2t}. \tag{E.41}$$

The solution to the homogeneous equation is given by Eq. (E.35). Note that

$$Q(t) = 2t^2 + 3e^{2t} \tag{E.42}$$

contains the term $\exp(2t)$, which, ignoring constant coefficients, is t^0 times the same term in the homogeneous solution, Eq. (E.35). Hence, $v(t)$ must contain a linear combination of $t\exp(2t)$ and all its linearly independent derivatives that are not contained in $u(t)$. Consequently, $v(t)$ has the form

$$v(t) = At^2 + Bt + C + Dte^{2t}. \tag{E.43}$$

Note that $\exp(2t)$ is not included in Eq. (E.43) because it is already included in $u(t)$. Substituting Eq. (E.43) into Eq. (E.41) and setting the coefficients of the various linearly independent terms equal to zero allows the determination of A, B, C and D. They are

$$A = 1, \qquad B = 3, \qquad C = \frac{7}{2}, \qquad D = 3, \tag{E.44}$$

and the particular solution is

$$v(t) = t^2 + 3t + \frac{7}{2} + 3te^{2t}. \tag{E.45}$$

Thus, the general solution to Eq. (E.41) is

$$y(t) = u(t) + v(t) = C_1e^t + C_2e^{2t} + t^2 + 3t + \frac{7}{2} + 3te^{2t}. \tag{E.46}$$

E.6 Secular Terms

Consider the following linear, inhomogeneous equation

$$\frac{d^2y}{dt^2} + \omega^2 y = \beta\cos\omega t, \tag{E.47}$$

where ω and β are parameters. The solution to the initial value problem

$$y(0) = A, \qquad \frac{dy(0)}{dt} = 0, \tag{E.48}$$

is

$$y(t) = A\cos\omega t + \left(\frac{\beta}{2\omega}\right) t\sin t. \tag{E.49}$$

Note that the first term on the right-hand side is periodic, while the second term is oscillatory, but has an increasing amplitude. The second expression is an example of a so-called *secular term*.

References

1. G. Birkhoff and G. C. Rota, *Ordinary Differential Equations* (Ginn, Boston, 1962).
2. E. A. Coddington and N. Levinson, *Theory of Ordinary Differential Equations* (McGraw-Hill, New York, 1955). See Chapters 1, 2, and 3.
3. W. Kaplan, *Advanced Calculus* (Addison-Wesley; Reading, MA; 1952). See Chapter 8.
4. E. A. Kraut, *Fundamentals of Mathematical Physics* (McGraw-Hill, New York, 1967). See Sections 6–18 and 6–21.
5. S. L. Ross, *Differential Equations* (Blaisdell; Waltham, MA; 1964). See Chapters 10 and 11.
6. M. Tenenbaum and H. Pillard, *Ordinary Differential Equations* (Harper and Row, New York, 1963). See Chapters 4, 11, and 12.

Appendix F

Lindstedt-Poincaré Perturbation Method

This appendix contains a brief outline of the Lindstedt-Poincaré [1, 2] perturbation method. It can be applied to construct uniformly valid expressions for the periodic solutions of second-order differential equations having the form

$$\ddot{x} + x = \epsilon F(x, \dot{x}), \tag{F.1}$$

where $\dot{x}$ denotes dx/dt, etc.; and ϵ is a small parameter, i.e.,

$$0 < \epsilon \ll 1. \tag{F.2}$$

A uniformly valid expression for this solution is one that can be written as

$$x(t, \epsilon) = \sum_{m=0}^{n} e^m x_m(t) + O(\epsilon^{n+1}), \tag{F.3}$$

where

$$|x_m(t)| < \infty; \quad m = 0, 1, 2 \ldots; \quad t > 0. \tag{F.4}$$

This method for calculating periodic solutions generally produces an asymptotic expansion in ϵ and has been the topic of a vast literature. In particular, the books by Stoker [3], Bellman [4], Nayfeh [5], Mickens [6], and Murdock [7], collectively provide the fundamental theoretical basis for this technique and also illustrate its many applications.

The general procedure is to substitute Eq. (F.3) into Eq. (F.1), expand the resulting expression in powers of ϵ, and set the coefficients of the difference powers of ϵ to zero. This procedure leads to a set of linear, inhomogeneous, second-order, differential equations for the $x_m(t)$, where $m = 0, 1, \ldots, n$; and these equations may be solved recursively. However, the straightforward application of the method will produce solutions having secular terms. A way to prevent this difficulty is to carry out the following procedure:

1) Introduce a new independent variable θ, i.e.,

$$\theta = \omega t, \tag{F.5}$$

where

$$\omega(\epsilon) = 1 + \omega_1\epsilon + \cdots + \epsilon^n\omega_n + O(\epsilon^{n+1}). \tag{F.6}$$

2) The new dependent variable is denoted by $x(\theta)$ and we assume that it can be represented as

$$x(\theta, \epsilon) = x_0(\theta) + \epsilon x_1(\theta) + \cdots + \epsilon^n x_n(\theta) + O(\epsilon^{n+1}). \tag{F.7}$$

Note that both the ω_m and $x_m(t)$, for $m = 1, 2, \ldots, n$, must be determined.

Introduce the following notation

$$x' \equiv \frac{dx}{d\theta}, \quad x'' \equiv \frac{d^2x}{d\theta^2},$$

$$F_x(x, \dot{x}) \equiv \frac{\partial F}{\partial x}, \quad F_{\dot{x}}(x, \dot{x}) = \frac{\partial F}{\partial x}.$$

3) Substitute Eq. (F.7) into Eq. (F.1), carry out the required expansions with respect to ϵ, and then set the coefficients of the various power of ϵ to zero; the following set of equations is obtained for x_m;

$$x_0'' + x_0 = 0 \tag{F.8}$$

$$x_1'' + x_1 = -2\omega_1 x_0'' + F(x_0, \dot{x}_0') \tag{F.9}$$

$$\begin{aligned} x_2'' + x_2 = &-2\omega_1 x_1'' - (\omega_1^2 + 2\omega_2)x_0'' \\ &+ F_x(x_0, x_0')x_1 + F_{\dot{x}}(x_0, x_0')(\omega_1 x_0' + x_1') \end{aligned} \tag{F.10}$$

$$\vdots \quad \vdots \quad \vdots$$

$$x_n'' + x_n = G_n(x_0, x_1, \ldots, x_{n-1}; x_0', x_1', \ldots, x_{n-1}'). \tag{F.11}$$

If $F(x, \dot{x})$ is a polynomial function, then G_n will also be a polynomial function of its variables.

4) The periodicity condition for $x(\theta)$ can be expressed as

$$x(\theta) = x(\theta + 2\pi), \tag{F.12}$$

and the corresponding condition for $x_m(\theta)$ are

$$x_m(\theta) = x_m(\theta + 2\pi). \tag{F.13}$$

If Eq. (F.7) is to be a periodic function of Eq. (F.1), then the right-hand sides of Eqs. (F.8) to (F.11) must not contain terms in either $\cos\theta$ or $\sin\theta$; otherwise, secular terms would exist and the perturbation solution will not be uniformly valid, i.e., one or more of the $x_m(t)$ might have the behavior

$$\lim_{t\to\infty} |x_m(t)| = \infty. \tag{F.14}$$

Therefore, if $x_m(\theta)$ is to be a periodic function, then, in general, two conditions must be satisfied at each step of the calculation. Thus, two "free parameters" are required. Examination of Eqs. (F.8) to (F.11) indicates that one of these constants is ω_m. The only other place where a second constant can be introduced is from the initial condition on x_{m-1} and this implies that the initial conditions should take the form

$$x(0) = A_0 + \epsilon A_1 + \epsilon^2 A_2 + \cdots + \epsilon^n A_n + O(\epsilon^{n+1}) \tag{F.15a}$$

$$x'(0) = 0, \tag{F.15b}$$

where the A_m are, *a priori*, unknown parameters. This means that for $m = 1$, the right-hand side of Eq. (F.9) has a term linear in ω_1 and another term nonlinear in A_0. These two terms are expressions involving $\cos\theta$ and $\sin\theta$, and by setting their respective coefficients to zero, not only are the secular terms eliminated, but both ω_1 and A_0 may be calculated. In a similar manner, for $m \geq 2$, the periodicity condition for $y_m(\theta)$ gives a pair of equations for ω_m and A_{m-1}. Again, the requirement of no secular terms in the solution for $y_m(\theta)$ allows both ω_m and A_{m-1} to be calculated. Thus, at any given stage (i.e., m value) in the procedure, the three quantities $(\omega_m, A_{m-1}, y_m(\theta))$ may be simultaneously determined. This means that the initial conditions are

$$\begin{cases} x_0(0) = A, & x_0'(0) = 0; \\ x_m(0) = A_m, & x_m'(0) = 0; \quad m = 1, 2, \ldots, n. \end{cases} \tag{F.16}$$

For the special cases where $F(x, \dot{x})$ either does not depend on $\dot{x}$ or is an even function of $\dot{x}$, then $x(\theta)$ can be chosen to be an even function by using the initial conditions

$$x(0) = A, \quad \dot{x}(0) = 0. \tag{F.17}$$

Therefore, $x(\theta)$ and $x_m(\theta)$ are even functions of θ and the right-hand sides of Eqs. (F.9) to (F.11) do not have terms in $\sin\theta$. For this case, only one parameter ω_m is required to ensure that there is no term in $\cos\theta$. This means that Eqs. (F.15) become

$$x(0) = A, \quad x'(0) = 0, \tag{F.18}$$

or

$$\begin{cases} x_0(0) = A, & x_0'(0) = 0, \\ x_m(0) = 0, & x_m'(0) = 0; \quad m = 1, 2, \ldots, n. \end{cases} \tag{F.19}$$

In summary, the $(n+1)$th approximation to the solution of Eq. (F.1) is

$$x(\theta, \epsilon) = \sum_{m=0}^{n} \epsilon^m x_m(\theta) + O(\epsilon^{n+1}),$$

where

$$\theta = \omega(\epsilon)t = \sum_{m=0}^{n} \omega_m \epsilon^m + O(\epsilon^{n+1}),$$

with $\omega_0 = 1$.

References

1. A. Lindstedt, *Astron. Nach.* **103**, 211 (1882).
2. H. Poincaré, *New Methods in Celestial Mechanics*, Vols. I, II, and III (English translation, NASA Reports TTF-450, -451, -452; 1967).
3. J. J. Stoker, *Nonlinear Vibrations in Mechanical and Electrical Systems* (Interscience, New York, 1950).
4. R. Bellman, *Perturbation Techniques in Mathematics, Physics and Engineering* (Holt, Rinehart and Winston; New York, 1966).
5. A. H. Nayfeh, *Perturbation Methods* (Wiley, New York, 1973).
6. R. E. Mickens, *Nonlinear Oscillations* (Cambridge University Press, New York, 1981).
7. J. A. Murdock, *Perturbations: Theory and Methods* (Wiley-Interscience, New York, 1991).

Appendix G

A Standard Averaging Method

The first published work on the method of averaging was the volume by Krylov and Bogoliubov [1]. This procedure has been extended and justified mathematically by Bogoliubov and Mitropolsky [2], and Minorsky [3]. Since then the method has become a standard technique for investigating nonlinear oscillator systems modeled by a differential equation taking the form

$$\ddot{x} + x = \epsilon F(x, \dot{x}), \quad 0 < \epsilon \ll 1, \tag{G.1}$$

where ϵ is a small parameter. The important feature of the method is that it not only allows the determination of steady-state periodic solutions, but also permits the determination of the transitory behavior of the system to a limit-cycle periodic solution. Excellent discussions on this procedure, along with many worked examples, are given in the books by Mickens [4] and Nayfeh [5]. A closely related technique was proposed by van der Pol [6] who studied the periodic oscillations of nonlinear vacuum tubes.

In this appendix, we provide a heuristic derivation of the first-approximation for the averaging method when applied to Eq. (G.1).

If $\epsilon = 0$, then Eq. (G.1) reduces to the linear equation

$$\ddot{x} + x = 0. \tag{G.2}$$

The general solution and its derivative are

$$x(t) = a\cos(t + \phi), \tag{G.3a}$$

$$\dot{x}(t) = -a\sin(t + \phi), \tag{G.3b}$$

where a and ϕ are arbitrary constants of integration.

Assume, for $0 < \epsilon \ll 1$, Eq. (G.1) has a solution that takes the form

$$x(t) = a(t)\cos[t + \phi(t)], \tag{G.4}$$

where $a(t) \equiv a(t, \epsilon)$ and $\phi(t) \equiv \phi(t, \epsilon)$ are both functions of t and ϵ. If we further assume that the derivative of $x(t)$ is

$$\dot{x}(t) = -a(t) \sin[t + \phi(t)], \tag{G.5}$$

then it follows from differentiating Eq. (G.4) that

$$\dot{x} = \dot{x} \cos\psi - a \sin\psi - a\dot{\phi} \sin\psi. \tag{G.6}$$

Therefore, for Eq. (G.5) to hold it must be that the following condition is satisfied

$$\dot{a} \cos\psi - a\dot{\phi} \sin\psi = 0, \tag{G.7}$$

where

$$\psi(t) = t + \phi(t). \tag{G.8}$$

If Eq. (G.4) is differentiated, we find

$$\ddot{x} = -\dot{a} \sin\psi - a \cos\psi - a\dot{\phi} \cos\psi. \tag{G.9}$$

Substituting Eqs. (G.4), (G.5) and (G.9) into Eq. (G.1) gives

$$\dot{a} \sin\psi + a\dot{\phi} \cos\psi = -\epsilon F(a \cos\psi, -a \sin\psi). \tag{G.10}$$

Since Eqs. (G.7) and (G.10) are linear in $\dot{a}$ and $\dot{\phi}$, they may be solved to obtain

$$\dot{a} = -\epsilon F(a \cos\psi, -a \sin\psi) \sin\psi, \tag{G.11a}$$

$$\dot{\phi} = -\left(\frac{\epsilon}{a}\right) F(a \cos\psi, -a \sin\psi) \cos\psi, \tag{G.11b}$$

$$\psi(t) = t + \phi(t). \tag{G.11c}$$

These expressions are the exact first-order differential equations for $a(t, \epsilon)$ and $\phi(t, \epsilon)$, when the solution and its derivative take the forms given by Eqs. (G.4) and (G.5). In general, these equations cannot be solved for $a(t, \epsilon)$ and $\phi(t, \epsilon)$. Therefore, an approximation method must be created such that the resulting equations can be solved for quantities approximating $a(t, \epsilon)$ and $\phi(t, \epsilon)$.

Inspection of Eqs. (G.11a) and (G.11b) shows that their right-hand sides are periodic in ψ with period 2π. Assume that Fourier expansions exist for $F \sin\psi$ and $F \cos\psi$, i.e.,

$$F(a \cos\psi, -a \sin\psi) \sin\psi = K_0(a) + \sum_{m=1}^{\infty} \Big[K_m(a) \cos(m\psi) + L_m(a) \sin(m\psi) \Big], \tag{G.12}$$

$$F(a\cos\psi, -a\sin\psi)\cos\psi = P_0(a) + \sum_{m=1}^{\infty}\Big[P_m(a)\cos(m\psi) + Q_m(a)\sin(m\psi)\Big], \quad \text{(G.13)}$$

where

$$K_0(a) = \left(\frac{1}{2\pi}\right)\int_0^{2\pi} F\sin\psi\, d\psi, \quad \text{(G.14a)}$$

$$K_m(a) = \left(\frac{1}{\pi}\right)\int_0^{2\pi} F\sin\psi\cos(m\psi)d\psi, \quad \text{(G.14b)}$$

$$L_m(a) = \left(\frac{1}{\pi}\right)\int_0^{2\pi} F\sin\psi\sin(m\psi)d\psi, \quad \text{(G.14c)}$$

$$P_0(a) = \left(\frac{1}{2\pi}\right)\int_0^{2\pi} F\cos\psi\, d\psi, \quad \text{(G.14d)}$$

$$P_m(a) = \left(\frac{1}{\pi}\right)\int_0^{2\pi} F\cos\psi\cos(m\psi)d\psi, \quad \text{(G.14e)}$$

$$Q_m(a) = \left(\frac{1}{\pi}\right)\int_0^{2\pi} F\cos\psi\sin(m\psi)d\psi. \quad \text{(G.14f)}$$

$$\text{(G.14g)}$$

With these relations, Eqs. (G.11a) and (G.11b) may be written

$$\dot{a} = -\epsilon K_0(a) - \epsilon\sum_{m=1}^{\infty}[K_m(a)\cos(m\psi) + L_m(a)\sin(m\psi)], \quad \text{(G.15a)}$$

$$\dot{\phi} = -\left(\frac{\epsilon}{a}\right)P_0(a) - \left(\frac{\epsilon}{a}\right)\sum_{m=1}^{\infty}[P_m(a)\cos(m\psi) + Q_m(a)\sin(m\psi)]. \quad \text{(G.15b)}$$

The first approximation of Krylov and Bogoliubov consists of neglecting all terms on the right-hand sides of Eqs. (G.15), i.e.,

$$\dot{a} = -\epsilon K_0(a), \quad \dot{\phi} = -\left(\frac{\epsilon}{a}\right)P_0(a). \quad \text{(G.16)}$$

Written out in full, we obtain the two relations

$$\dot{a} = -\left(\frac{\epsilon}{2\pi}\right)\int_0^{2\pi} F(a\cos\psi, -a\sin\psi)\sin\psi\, d\psi, \quad \text{(G.17a)}$$

$$\dot{\phi} = -\left(\frac{\epsilon}{2\pi a}\right)\int_0^{2\pi} F(a\cos\psi, -a\sin\psi)\cos\psi\, d\psi. \quad \text{(G.17b)}$$

Note that the right-sides are both functions only of a. Therefore, the general method of first-order averaging is to solve Eq. (G.17a) and substitute this value for a into Eq. (G.17b) and solve the resulting expression for ϕ.

In summary, the first approximation of Krylov and Bogoliubov for the equation

$$\ddot{x} + x = \epsilon F(x, \dot{x}), \quad 0 < \epsilon \ll 1,$$

is the expression

$$x(t, \epsilon) = a(t, \epsilon)\cos[t + \phi(t, \epsilon)],$$

where $a(t, \epsilon)$ and $\phi(t, \epsilon)$ are solutions to Eqs. (G.17). This procedure is also called the method of first-order averaging [1, 2, 3] and the method of slowly varying amplitude and phase [4, 7].

References

1. N. Krylov and N. Bogoliubov, *Introduction to Nonlinear Mechanics* (Princeton University Press; Princeton, NJ; 1943).
2. N. N. Bogoliubov and Y. A. Mitropolsky, *Asymptotical Methods in the Theory of Nonlinear Oscillations* (Hindustan Publishing Co.; Delhi, India; 1963).
3. N. Minorsky, *Introduction to Nonlinear Mechanics* (J. W. Edwards; Ann Arbor, MI; 1947).
4. R. E. Mickens, *Nonlinear Oscillations* (Cambridge University Press, New York, 1981).
5. A. H. Nayfeh, *Perturbation Methods* (Wiley, New York, 1973).
6. B. van der Pol, *Philosophical Magazine* **43**, 700 (1926).
7. N. W. McLachlan, *Ordinary Nonlinear Differential Equations in Engineering and Physical Sciences* (Oxford University Press, London, 1950).

Appendix H

Discrete Models of Two TNL Oscillators

Truly nonlinear (TNL) differential equations do not in general have exact solutions expressible as a finite representation of the elementary functions [1]. Therefore, numerical integration procedures must be used to determine explicit solutions that may then be compared to the results of analytical approximations. This appendix constructs finite difference schemes for two versions of the cube-root equation [2]. These schemes are based on the nonstandard finite (NSFD) methodology of Mickens [3, 4, 5, 6]. Reference [3] provides a broad introduction to the background required to both understand and apply this numerical integration procedure.

H.1 NSFD Rules [3, 6]

The NSFD methodology is based on two requirements [3]. The first is that the discretization of the first derivative takes the general form

$$\frac{dx}{dt} \rightarrow \frac{x_{k+1} - x_k}{\phi(h)}, \tag{H.1}$$

where

$$x(t) \rightarrow x(t_k), \quad t \rightarrow t_k = hk, \quad h = \Delta t, \tag{H.2}$$

and the *denominator function*, $\phi(h)$, has the property

$$\phi(h) = h + O(h^2). \tag{H.3}$$

The second requirement is that functions of the dependent variable x are discretized nonlocally on the k-computational grid. In general, this means that $f(x)$ has a representation for which the x's have terms involving k, $k-1$, etc. For example, the following are possible discretizations [3, 4, 6]:

$$x^2 \rightarrow \begin{cases} x_k x_{k+1}, & \text{1st-order ODE;} \\ \left(\dfrac{x_{k+1} + x_k + x_{k-1}}{3}\right) x_k, & \text{2nd-order ODE.} \end{cases}$$

$$x^3 \to \begin{cases} x_k^2 x_{k+1} + x_k x_{k+1}^2, & \text{1st-order ODE;} \\ \left(\dfrac{x_{k+1} + x_{k-1}}{2}\right) x_k^2, & \text{2nd-order ODE.} \end{cases}$$

H.2 Discrete Energy Function [4, 5]

Consider a conservative nonlinear oscillator modeled by the differential equation

$$\ddot{x} + g(x) = 0. \tag{H.4}$$

This equation has the following first-integral

$$H(x, \dot{x}) \equiv \frac{(\dot{x})^2}{2} + V(x) = \text{constant}, \tag{H.5}$$

where

$$V(x) = \int^x g(z)dz. \tag{H.6}$$

Within the NSFD methodology, a discretization of the first-integral, $H(x, \dot{x})$, should have the property of being invariant under the interchange

$$k \leftrightarrow (k-1), \tag{H.7}$$

i.e., if $H_k = H(x_k, x_{k-1})$, then

$$H(x_k, x_{k-1}) = H(x_{k-1}, x_k). \tag{H.8}$$

From a knowledge of $H(x_k, x_{k-1})$, the second-order difference equation corresponding to Eq. (H.4) can be constructed from the relation

$$\Delta H(x_k, x_{k-1}) = 0, \tag{H.9}$$

where

$$\Delta f_k \equiv f_{k+1} - f_k.$$

Therefore, using this definition of the Δ operator, we have

$$H(x_{k+1}, x_k) = H(x_k, x_{k-1}). \tag{H.10}$$

H.3 Cube-Root Equation [2]

The cube-root TNL oscillator differential equation is

$$\ddot{x} + x^{1/3} = 0. \tag{H.11}$$

Its first-integral is

$$H(x, \dot{x}) = \left(\frac{1}{2}\right)(\dot{x})^2 + \left(\frac{3}{4}\right) x^{4/3} = \text{constant}. \tag{H.12}$$

A discretization of $H(x, \dot{x})$ that satisfies the condition given in Eq. (H.7) is

$$H(x_k, x_{k-1}) = \left(\frac{1}{2}\right)\left(\frac{x_k - x_{k-1}}{\phi}\right)^2 + \left(\frac{3}{4}\right) x_k^{2/3} x_{k-1}^{2/3}. \tag{H.13}$$

The calculation of $\Delta H(x_k, x_{k-1}) = 0$ requires the evaluation of two terms. The first is

$$\begin{aligned}\Delta(x_k - x_{k-1})^2 &= \Delta(x_k^2 - 2x_k x_{k-1} + x_{k-1}^2)\\ &= (x_{k+1}^2 - x_k^2) - 2(x_{k+1}x_k - x_k x_{k-1}) + (x_k^2 - x_{k-1}^2)\\ &= (x_{k+1} - 2x_k + x_{k-1})(x_{k+1} - x_{k-1}),\end{aligned} \tag{H.14}$$

while the second is

$$\begin{aligned}\Delta(x_k^{2/3} x_{k-1}^{2/3}) &= x_{k+1}^{2/3} x_k^{2/3} - x_k^{2/3} x_{k-1}^{2/3}\\ &= (x_{k+1}^{2/3} - x_{k-1}^{2/3}) x_k^{2/3}.\end{aligned} \tag{H.15}$$

Placing these results in $\Delta H(x_k, x_{k-1}) = 0$ gives

$$\frac{x_{k+1} - 2x_k + x_{k-1}}{\phi^2} + \left(\frac{2}{3}\right)\left(\frac{x_{k+1}^{2/3} - x_{k-1}^{2/3}}{x_{k+1} - x_{k-1}}\right) x_k^{2/3} = 0.$$

Using the relation

$$\frac{a^2 - b^2}{a^3 - b^3} = \frac{(a+b)(a-b)}{(a-b)(a^2 + ab + b^2)} = \frac{a+b}{a^2 + ab + b^2},$$

with

$$a = x_{k+1}^{1/3}, \quad b = x_{k-1}^{1/3},$$

we find

$$\frac{x_{k+1} - 2x_k + x_{k-1}}{\phi^2} + \left[\frac{\left(x_{k+1}^{1/3} + x_{k-1}^{1/3}\right)/2}{\left(x_{k+1}^{2/3} + x_{k+1}^{1/3} x_{k-1}^{1/3} + x_{k-1}^{2/3}\right)/3}\right] x_k^{2/3} = 0,$$

or

$$\left(\frac{x_{k+1} - 2x_k + x_{k-1}}{\phi^2}\right) + \left[\frac{3x_k^{2/3}}{x_{k+1}^{2/3} + x_{k+1}^{1/3}x_{k-1}^{1/3} + x_{k-1}^{2/3}}\right]\left(\frac{x_{k+1}^{1/3} + x_{k-1}^{1/3}}{2}\right) = 0. \tag{H.16}$$

This is the NSFD scheme for Eq. (H.11).

Other NSFD schemes may be constructed for the cube-root differential equation. If this equation is written as a system of two coupled first-order equations

$$\frac{dx}{dt} = y, \quad \frac{dy}{dt} = -x^{1/3}, \tag{H.17}$$

then we obtain the discretizations:

NSFD-1

$$\frac{x_{k+1} - x_k}{\phi} = y_k, \quad \frac{y_{k+1} - y_k}{\phi} = -x_{k+1}^{1/3}, \tag{H.18}$$

or

$$x_{k+1} = x_k + \phi y_k, \quad y_{k+1} = y_k - \phi(x_k + \phi y_k)^{1/3}; \tag{H.19}$$

NSFD-2

$$\frac{x_{k+1} - x_k}{\phi} = y_{k+1}, \quad \frac{y_{k+1} - y_k}{\phi} = -x_k^{1/3}, \tag{H.20}$$

or

$$x_{k+1} = x_k + \phi y_k - \phi^2 x_k^{1/3}, \quad y_{k+1} = y_k - \phi y_k^{1/3}. \tag{H.21}$$

Numerical experiments were performed using these three NSFD schemes. For these simulations, ϕ was selected to be h, i.e., $\phi(h) = h$. The numerical results, in each case, produced solutions that oscillated with the expected constant amplitudes. However, two other standard schemes give results inconsistent with the known behavior of the cubic oscillator. A forward-Euler discretization produced numerical solutions that oscillated with increasing amplitude, while the standard MATLAB one-step ODE solver `ode45` gave numerical solutions with decreasing oscillatory amplitudes. These results clearly demonstrate the overall dynamic consistency of the NSFD methodology as compared with many of the standard numerical integration methods [2, 19].

H.4 Cube-Root/van der Pol Equation

This TNL oscillator equation is

$$\ddot{x} + x^{1/3} = \epsilon(1 - x^2)\dot{x}. \tag{H.22}$$

We now suggest a possible NSFD scheme for this differential equation. We begin with the observation that the van der Pol oscillator

$$\ddot{x} + x = \epsilon(1 - x^2)\dot{x}, \tag{H.23}$$

has a NSFD representation given by the following expression (see Section 5.4 in Mickens [3]),

$$\frac{x_{k+1} - 2x_k + x_{k-1}}{4\sin^2\left(\frac{h}{2}\right)} + x_k = \epsilon(1 - x_k^2)\left[\frac{x_k - x_{k-1}}{\left(\frac{4\sin^2(h/2)}{h}\right)}\right]. \tag{H.24}$$

Based on the discretizations, given in Eqs. (H.16) and (H.24), we take the following result for the NSFD scheme of the cube-root/van der Pol equation

$$\frac{x_{k+1} - 2x_k + x_{k-1}}{4\sin^2\left(\frac{h}{2}\right)} + \left[\frac{3x_k^{2/3}}{x_{k+1}^{2/3} + x_{k+1}^{1/3}x_{k-1}^{1/3} + x_{k-1}^{2/3}}\right]\left[\frac{x_{k+1}^{1/3} + x_{k-1}^{1/3}}{2}\right]$$
$$= \epsilon(1 - x_k^2)\left[\frac{x_k - x_{k-1}}{\left(\frac{4\sin^2(h/2)}{h}\right)}\right]. \tag{H.25}$$

Comparing this structure with

$$\ddot{x} + x^{1/3} = \epsilon(1 - x^2)\dot{x},$$

the following conclusions may be reached regarding this discretization:

(i) The first-order derivative, on the right-hand side of the differential equation has the NSFD form

$$\dot{x} \to \frac{x_k - x_{k-1}}{\left[\frac{4\sin^2(h/2)}{h}\right]};$$

this corresponds to a backward-Euler scheme having the denominator function

$$\phi_1(h) = \frac{4\sin^2(h/2)}{h}, \tag{H.26}$$

i.e.,

$$\dot{x} \to \frac{x_k - x_{k-1}}{\phi_1(h)}, \tag{H.27}$$

where

$$\phi_1(h) = h + O(h^2)$$

(ii) The quadratic function $(1 - x^2)$ has a local representation, i.e.,

$$(1 - x^2) \to (1 - x_k^2). \tag{H.28}$$

(iii) The second-derivative, $\ddot{x}$, is a generalization of the central-difference scheme, i.e.,

$$\ddot{x} \to \frac{x_{k+1} - 2x_k + x_{k-1}}{h^2},$$

and corresponds to the replacement

$$\ddot{x} \to \frac{x_{k+1} - 2x_k + x_{k-1}}{[\phi_2(h)]^2}, \tag{H.29}$$

where the denominator function is

$$\phi_2(h) = 2 \sin\left(\frac{h}{2}\right). \tag{H.30}$$

(iv) The discretization of $x^{1/3}$ is the complex expression

$$x^{1/3} \to \left(\frac{3x_k^{2/3}}{x_{k+1}^{2/3} + x_{k+1}^{1/3}x_{k-1}^{1/3} + x_{k-1}^{2/3}}\right)\left(\frac{x_{k+1}^{1/3} + x_{k-1}^{1/3}}{2}\right). \tag{H.31}$$

Examination of the right-hand side of this expression indicates that the "$x^{1/3}$" term is averaged over the two grid points, $t = t_{k-1}$ and t_{k+1}, and this quantity is multiplied by a factor that is essentially "one," in the sense that in the limit where $k \to \infty$, $h \to 0$, $t_k = t =$ constant, then its value becomes one.

References

1. D. Zwillinger, *Handbook of Differential Equations* (Academic Press, Boston, 1989).
2. M. Ehrhardt and R. E. Mickens, *Neural Parallel and Scientific Computations* **16**, 179 (2008).
3. R. E. Mickens, *Nonstandard Finite Difference Models of Differential Equations* (World Scientific, Singapore, 1994).
4. R. E. Mickens, *Journal of Difference Equations and Applications* **2**, 185 (1996).
5. R. Anguelov and J. M.-S. Lubuma, *Numerical Methods for Partial Differential Equations* **17**, 518 (2001).
6. R. E. Mickens, *Journal of Difference Equations and Applications* **11**, 645 (2005).

Bibliography

Linear Analysis and Differential Equations

Agnew, R. P., *Differential Equations* (McGraw-Hill, New York, 1960).

Apostol, T. M., *Mathematical Analysis* (Addison-Wesley, Reading, MA, 1957).

Birkhoff, G. and G. Rota, *Ordinary Differential Equations* (Ginn, Boston, 1962).

Boyce, W. and R. Diprima, *Elementary Differential Equations and Boundary Value Problems* (Wiley, New York, 1969), 2nd ed.

Churchill, R. V., *Fourier Series and Boundary Value Problems* (McGraw-Hill, New York, 1941).

Coddington, E. A., *An Introduction to Ordinary Differential Equations* (Prentice-Hall, Englewood Cliffs, NJ, 1961).

Courant, R. and D. Hilbert, *Methods of Mathematical Physics, Vol. I* (Interscience, New York, 1953).

Ford, L. R., *Differential Equations* (McGraw-Hill, New York, 1955).

Jeffreys, H. and B. Jeffreys, *Methods of Mathematical Physics* (Cambridge University Press, Cambridge, 1962).

Kaplan, W., *Ordinary Differential Equations* (Addison-Wesley, Reading, MA, 1958).

Mickens, R. E., *Matheamtical Methods for the Natural and Engineering Sciences* (World Scientific, Singapore, 2004).

Murphy, G., *Ordinary Differential Equations and Their Solution* (Van Nostrand Reinhold, New York, 1960).

Rainville, E. D., *Elementary Differential Equations* (Macmillan, New York, 1958).

Sneddon, I. N., *Elements of Partial Differential Equations* (McGraw-Hill, New York, 1957).

Widder, D. V., *Advanced Calculus* (Prentice-Hall, Englewood Cliffs, NJ, 1960).

Yosida, K., *Lectures on Differential and Integral Equations* (Interscience, New York, 1960).

Nonlinear Analysis and Regular Perturbation Methods

Aggarwal, J. K., *Notes on Nonlinear Systems* (Van Nostrand Reinhold, New York, 1972).

Bellman, R., *Stability Theory of Differential Equations* (McGraw-Hill, New York, 1954).

Bellman, R., *Perturbation Techniques in Mathematics, Physics and Engineering* (Holt, Rinehart and Winston, New York, 1964).

Bush, A. W., *Perturbation Methods for Engineers and Scientists* (CRC Press, Boca Raton, FL, 1992).

Bogoliubov, N. N., Ju. A. Mitropoliskii, and A. M. Samoilenko, *Methods of Accelerated Convergence in Nonlinear Mechanics* (Springer-Verlag, New York, 1976).

Cesari, L., *Asymptotic Behavior and Stability Problems in Ordinary Differential Equations* (Springer-Verlag, Berlin, 1959).

Coddington, E. A. and N. Levinson, *Theory of Ordinary Differential Equations* (McGraw-Hill, New York, 1955).

Cole, J. D., *Perturbation Methods in Applied Mathematics* (Blaisdell, Waltham, MA, 1968).

Davies, H. T., *Introduction to Nonlinear Differential and Integral Equations* (Dover, New York, 1962).

Davies, T. V. and E. M. James, *Nonlinear Differential Equations* (Addison-Wesley, Reading, MA, 1966).

de Figueiredo, R. P., *Contribution to the Theory of Certain Nonlinear Differential Equations* (Lisbon, 1960).

Fedoryuk, M. V., *Asymptotic Analysis* (Springer-Verlag, Berlin, 1993).

Giacaglia, G. E., *Perturbation Methods in Nonlinear Systems* (Springer-Verlag, New York, 1972).

Glendinning, P., *Stability, Instability and Chaos* (Cambridge University Press, Cambridge, 1994).

Greenspan, D., *Theory and Solution of Ordinary Differential Equations* (Macmillan, New York, 1960).

Hurewicz, E., *Lectures on Ordinary Differential Equations* (Wiley, New York, 1958).

Ince, E. L., *Ordinary Differential Equations* (Dover, New York, 1956).

Kato, T., *A Short Introduction to Perturbation Theory for Linear Operators* (Springer-Verlag, New York, 1982).

Lefschetz, S., *Differential Equations, Geometric Theory* (Interscience, New York, 1957).

Mickens, R. E., *Nonlinear Oscillations* (Cambridge University Press, New York, 1981).

Murdock, J. A., *Perturbations: Theory and Methods* (Wiley-Interscience, New York, 1991).

Nayfeh, A. H., *Perturbation Methods* (Wiley, New York, 1973).

Nemytskii, V. and V. Stepanov, *Qualitative Theory of Differential Equations* (Princeton University Press, Princeton, NJ, 1959).

Perko, Lawrence, *Differential Equations and Dynamical Systems* (Springer-Verlag, New York, 1991).

Rand, R. H. and D. Armbruster, *Perturbation Methods, Bifurcation Theory and Computer Algebra* (Springer-Verlag, New York, 1987).

Saaty, T. L. and J. Bram, *Nonlinear Mathematics* (McGraw-Hill, New York, 1964).

Sansone, G. and R. Conti, *Nonlinear Mathematics* (McGraw-Hill, New York, 1964).

Sears, W. R., *Small Perturbation Theory* (Princeton University Press, Princeton, NJ, 1960).

Simmonds, J. G. and J. E. Mann, Jr., *A First Look at Perturbation Theory* (Dover, New York, 1997), 2nd ed.

Struble, R. A., *Nonlinear Differential Equations* (McGraw-Hill, New York, 1962).

Nonlinear Oscillations

Andronov, A. A. and C. E. Chaikin, *Theory of Oscillations* (Princeton University Press, Princeton, NJ, 1949).

Andronov, A. A., A. A. Vitt and S. E. Khaikin, *Theory of Oscillators* (Addison-Wesley, Reading, MA, 1966).

Bobylev, N. A., Y. M. Burman and S. K. Korovin, *Approximation Procedures in Nonlinear Oscillation Theory* (Walter de Gruyter, Berlin, 1994).

Bogoliubov, N. N. and Y. A. Mitropolsky, *Asymptotic Methods in the Theory of Non-linear Oscillations* (Hindustan Publishing, Delhi, 1961).

Burton, T. A., *Stability and Periodic Solutions of Ordinary and Functional Differential Equations* (Dover, New York, 2005).

Butenin, N. N., *Elements of the Theory of Nonlinear Oscillations* (Blaisdell,

New York, 1965).

Cunningham, W. J., *Introduction to Nonlinear Analysis* (McGraw-Hill, New York, 1958).

Dinca, F. and C. Teodosiu, *Nonlinear and Random Vibrations* (Academic, New York, 1973).

Farkas, M., *Periodic Motions* (Springer-Verlag, New York, 1994).

Haag, J., *Oscillatory Motions* (Wadsworth, Belmont, CA, 1962).

Hale, J. K., *Oscillations in Nonlinear Systems* (McGraw-Hill, New York, 1963).

Hartog, J. P. D., *Mechanical Vibrations* (McGraw-Hill, New York, 1956), 4th ed.

Hayashi, C., *Forced Oscillations in Nonlinear Systems* (Nippon Printing, Osaka, Japan, 1953).

Hayashi, C., *Nonlinear Oscillations in Physical Systems* (McGraw-Hill, New York, 1964).

Hilborn, R. C., *Chaos and Nonlinear Dynamics* (Oxford University Press, New York, 1994).

Krasnosel'skii, M. A., V. Sh. Burd, and Yu. S. Kolesov, *Nonlinear Almost Periodic Oscillations* (Wiley, New York, 1973).

Kryloff, N. and N. Bogoliubov, *Introduction to Nonlinear Mechanics* (Princeton University Press, Princeton, NJ, 1943).

McLachlan, N. W., *Ordinary Nonlinear Differential Equations in Engineering and Physical Science* (Oxford, London, 1950).

McLachlan, N. W., *Theory of Vibrations* (Dover, New York, 1951).

Mickens, R. E., *Nonlinear Oscillations* (Cambridge University Press, New York, 1981).

Mickens, R. E., *Oscillations in Planar Dynamic Systems* (World Scientific, Singapore, 1996).

Minorsky, N., *Nonlinear Oscillation* (Van Nostrand Reinhold, Princeton, NJ, 1962).

Mitropolsky, Iu. A., *Nonstationary Processes in Nonlinear Oscillatory Systems.* Air Tech. Intelligence Transl. ATIC-270579, F-TS-9085/V.

Nayfeh, A. H., *Method of Formal Forms* (Wiley-Interscience, New York, 1993).

Schmidt, G. and A. Tondl, *Non-Linear Vibrations* (Cambridge University Press, Cambridge, 1986).

Stoker, J. J., *Nonlinear Vibrations* (Interscience, New York, 1950).

Urabe, M., *Nonlinear Autonomous Oscillations* (Academic, New York, 1967).

Wiggins, S., *Global Bifurcations and Chaos* (Springer-Verlag, New York, 1988).

Applications

Ames, W. F., *Nonlinear Ordinary Differential Equations in Transport Processes* (Academic, New York, 1968).

Beltrami, E., *Mathematics for Dynamic Modeling* (Academic, Boston, 1987).

Blaquiere, A., *Nonlinear System Analysis* (Academic, New York, 1966).

Edelstein-Keshet, L., *Mathematical Models in Biology* (McGraw-Hill, New York, 1987).

Gray, P. and S. K. Scott, *Chemical Oscillations and Instabilities* (Clarendon Press, Oxford, 1990).

Hughes, W. L., *Nonlinear Electrical Networks* (Ronald Press, New York, 1960).

Ku, Yü-hsiu, *Analysis and Control of Nonlinear Systems* (Ronald Press, New York, 1958).

Meirovitch, L., *Elements of Vibration Analysis* (McGraw-Hill, New York, 1975).

Minorsky, N., *Introduction to Nonlinear Mechanics* (J. W. Edwards, Ann Arbor, MI, 1947).

Murray, J. D., *Mathematical Biology* (Springer-Verlag, Berlin, 1989).

Pavlidis, T., *Biological Oscillators* (Academic, New York, 1973).

Perlmutter, A. and L. F. Scott (editors): *The Significance of Nonlinearity in the Natural Sciences* (Plenum, New York, 1977).

Pipes, L. A., *Operational Methods in Nonlinear Mechanics* (Dover, New York, 1965).

Pipes, L. A. and L. R. Harvill, *Applied Mathematics for Engineers and Physicists* (McGraw-Hill, New York, 1970).

Poincaré, H., *New Methods in Celestial Mechanics, Vols. I–III* (English translation), NASA TTF-450, 1967.

Scott, S. K., *Chemical Chaos* (Clarendon Press, Oxford, 1991).

Siljak, D., *Nonlinear Systems: The Parameter Analysis and Design* (Wiley, New York, 1969).

Strogatz, S. H., *Nonlinear Dynamics and Chaos with Application to Physics, Biology, Chemistry, and Engineering* (Addison-Wesley, Reading, MA, 1994).

Timoshenko, S., *Vibration Problems in Engineering* (Van Nostrand Reinhold, Princeton, NJ, 1937), 2nd ed.

Tu, P. N. V., *Dynamical Systems with Applications in Economics and Biology* (Springer-Verlag, Berlin, 1994), 2nd ed.
Van Dyke, M., *Perturbation Methods in Fluid Mechanics* (Academic, New York, 1964).

Qualitative Methods

Andronov, A. A., E. A. Leontovich, I. I. Gordon, and A. G. Maiser, *Qualitative Theory of Second-Order Dynamic Systems* (Wiley, New York, 1973, Israel Program for Scientific Translations).
Edelstein-Keshet, *Mathematical Models in Biology* (McGraw-Hill, New York, 1988). See Chapter 5.
Humi, M. and W. Miller, *Second Course in Ordinary Differential Equations for Scientists and Engineers* (Springer-Verlag, New York, 1988). See Chapter 8.
Liu, J. H., *A First Course in the Qualitative Theory of Differential Equations* (Pearson, Saddle River, NJ, 2003).
Martin, M., *Differential Equations and Their Applications*, (Springer-Verlag, New York, 1993, 4th edition). See Chapter 4.
Minorsky, N., *Nonlinear Oscillations* (Van Nostrand; Princeton, NJ; 1962). See Chapters 3 and 14.
Nemytskii V. and V. V. Stepanov, *Qualitative Theory of Differential Equations* (Princeton University Press; Princeton, NJ; 1960). See pp. 133–134.
Segel, L. A., editor, *Mathematical Models in Molecular and Cellular Biology* (Cambridge University Press, Cambridge, 1980). See Appendix A.3.
Whittaker, E. T., *Advanced Dynamics* (Cambridge University Press, London, 1937).

Selected Publications of R. E. Mickens on Truly Nonlinear Oscillations

1. "Comments on the Method of Harmonic Balance," *Journal of Sound and Vibration* **94**, 456 (1984).
2. "Approximate Analytic Solutions for Singular Nonlinear Oscillators," *Journal of Sound and Vibration* **96**, 277 (1984).
3. "Construction of Approximate Analytic Solutions to a New Class of Nonlinear Oscillator Equation," with K. Oyedeji, *Journal of Sound and Vibration* **102**, 579 (1985).
4. "A Generalization of the Method of Harmonic Balance," *Journal of Sound and Vibration* **111**, 515 (1986).

5. "Analysis of the Damped Pendulum," *Journal of Sound and Vibration* **115**, 375 (1987).
6. "Iteration Procedure for Determining Approximate Solutions to Non-Linear Oscillator Equations," *Journal of Sound and Vibration* **116**, 185 (1987).
7. "Application of Generalized Harmonic Balance to an Anti-Symmetric Quadratic Nonlinear Oscillator," with M. Mixon, *Journal of Sound and Vibration* **159**, 546 (1992).
8. "Harmonic Balance: Comparison of Equation of Motion and Energy Methods," with S. Hiamang, *Journal of Sound and Vibration* **164**, 179 (1993).
9. "Exact Solution to the Anti-Symmetric Constant Force Oscillator Equation," with T. Lipscomb, *Journal of Sound and Vibration* **169**, 138 (1994).
10. "Fourier Analysis of a Rational Harmonic Balance Approximation for Periodic Solutions," with D. Semwogerere, *Journal of Sound and Vibration* **195**, 528–530 (1996).
11. "A Phase-Space Analysis of a Nonlinear Oscillator Equation," with D. Semwogerere, *Journal of Sound and Vibration* **204**, 556–559 (1997).
12. "Regulation of Singular ODE's Modeling Oscillating Systems," *Journal of Sound and Vibration* **208**, 345–348 (1997).
13. "Periodic Solutions of the Relativistic Harmonic Oscillator," *Journal of Sound and Vibration* **212**, 905–908 (1998).
14. "Comment on a paper by M. S. Sarma and B. N. Rao – 'A Rational Harmonic Balance Approximation for the Duffing Equation of Mixed Parity'," *Journal of Sound and Vibration,* **216**, 187–189 (1998).
15. "Generalization of the Senator-Bapat Method to Systems Having Limit-Cycles," *Journal of Sound and Vibration* **224**, 167–171 (1999).
16. "Generalized Harmonic Oscillators," *Journal of Sound and Vibration* **236**, 730–732 (2000).
17. "Mathematical and Numerical Study of the Duffing-Harmonic Oscillator," *Journal of Sound and Vibration* **244**, 563–567 (2001).
18. "Oscillations in a $x^{4/3}$ Potential," *Journal of Sound and Vibration* **246**, 375–378 (2001).
19. "Generalized Harmonic Balance/Numerical Method for Determining Analytical Approximations to the Periodic Solutions of the $x^{4/3}$ Potential," with Karega Cooper, *Journal of Sound and Vibration* **250**, 951–954 (2002).
20. "A Study of Nonlinear Oscillations in Systems Having Non-Polynomial

Elastic Force Functions," *Recent Research Developments in Sound and Vibration* **1**, 241–251 (2002).

21. "Generalized Harmonic Oscillators: Velocity Dependent Frequencies," Conference Proceedings in CD-ROM format, 2001 *ASME DETC and CIE Conference* (The American Society of Mechanical Engineers; New York, 2001); paper DETC2001/VIB-21417.
22. "Analysis of Nonlinear Oscillators Having Nonpolynomial Elastic Terms," *Journal of Sound and Vibration* **255**, 789–792 (2002).
23. "Fourier Representations for Periodic Solutions of Odd-Parity Systems," *Journal of Sound and Vibration* **258**, 398–401 (2002).
24. "Fractional van der Pol Equations," *Journal of Sound and Vibration* **259**, 457–460 (2003).
25. "A Combined Equivalent Linearization and Averaging Perturbation Method for Nonlinear Oscillator Equations," *Journal of Sound and Vibration* **264**, 1195–1200 (2003).
26. "Mathematical Analysis of the Simple Harmonic Oscillator with Fractional Damping," *Journal of Sound and Vibration* **268**, 839–842 (2003), with K. O. Oyedeji and S. A. Rucker.
27. "A New Perturbation Method for Oscillatory Systems," *Conference Proceedings in CD-ROM Format,* 2003 *ASME DETC and CIE Conferences* (The American Society of Mechanical Engineers; New York, 2003); paper DETC03/VIB-48567.
28. "Preliminary Analytical and Numerical Investigations of a van der Pol Type Oscillator Having Discontinuous Dependence on the Velocity, *Journal of Sound and Vibration* **279**, 519–523 (2005), with K. Oyedeji and S. A. Rucker.
29. "A Pertrubation Method for Truly Nonlinear Oscillator Differential Equations," in G. S. Ladde, N. G. Medhin, and M. Sambandham (editors), *Proceedings of Dynamic Systems and Applications* **4**, 302–311 (2004), with S. A. Rucker.
30. "A Generalized Iteration Procedure for Calculating Approximations to Periodic Solutions of 'Truly Nonlinear Oscillators'," *Journal of Sound and Vibration* **287**, 1045–1051 (2005).
31. "Calculation of Analytic Approximations to the Periodic Solutions of a 'Truly Nonlinear' Oscillator Equation," Item DETC 2005–84474, CD-ROM Format, *Conference Proceedings of the* 2005 *American Society of Mechanical Engineering (ASME), IDETC and CIE* (ASME; September 24–28, 2005; Long Beach, CA).
32. "Iteration Method Solutions for Conservative and Limit-Cycle $x^{1/3}$

Force Oscillators," *Journal of Sound and Vibration* **292**, 964–968 (2006).
33. "Harmonic Balance and Iteration Calculations of Periodic Solutions to $\ddot{y} + y^{-1} = 0$," *Journal of Sound and Vibration* **306**, 968–972 (2007).
34. "Discrete Models for the Cube-Root Differential Equation," with M. Ehrhardt *Neural, Parallel and Scientific Computations* **16** (2008), 179–188.
35. "Exact and Approximate Values of the Period for a Truly Nonlinear Oscillator: $\ddot{x} + x + x^{1/3} = 0$," *Advances in Applied Mathematics and Mechanics* **1** (2009), 383–390.

Index